APPEARANCES

ISBN-10: 1463520166
EAN-13: 9781463520168
Library of Congress Control Number: 2011908832

Createspace, North Charleston, SC

APPEARANCES

Genetic Mythology and Cosmic Instincts

Laurie McRobert

For my grandchildren
Amanda, Dylan, Emily, Alexandra, Alex, and Natalie

CONTENTS

PREFACE

Like most people I am fascinated by the question, "Why am I here on a tiny blue dot, our earth, floating in a galaxy in an unfathomable ever-expanding space?" As a six-year-old I would try very hard to imagine how Jack and his magic beanstalk worked. I searched the heavens to see whether there might be a cord dangling down from space in the hopes of climbing it myself. As a more sophisticated ten-year-old I can remember taking a flashlight and a book outside to try to find, in the heavens, the constellations I was reading about. Graduating to a telescope to stargaze as an adult still found me searching the heavens for something more and questioning our human relationship with the cosmos.

My fascination with the cosmos and all its inherent mysteries has culminated in the writing of this book. It was inspired many years ago by my father's telling me the story of a White Lady who appeared to him when he was a teenager. She just rose out of the earth, naked, and stood still, with long flowing hair surrounding her, and then slowly disappeared. As he was just a farm boy, his story ruminated in my mind and fed my curiosity. Many others such as Boethius and Teilhard de Chardin wrote about such feminine appearances.[1] The important philosophers, poets, and mystics that I write about in this book had stories to tell about their personal experiences with appearances. Yet the human mind, unable to cope with this mystery, either sloughed it off as nonsense or chose to ignore it entirely.

So it was that after years of germinating in my mind, I came to write this book, which invites people to reflect about who they are in the greater scheme of things and why unexplained appearances—from UFOs to biblical appearances, such as Christ's appearances after his Crucifixion—play such a vital role in shaping the destiny of humankind. To write about appearances and the cosmos required me to penetrate regions of scholarly fringes where other philosophical scholars feared to tread. But I remain a philosopher and feel strongly about the need to push the frontiers of science and technology to new borders of space and time. The fact that ethereal appearances have stimulated human curiosity and inspired us to explore the heavens needs to be addressed. As our space programs struggle to push forward in the human quest to explore the cosmos, we find ourselves bumping up against the firewall of Einsteinian ideas of spacetime. These ideas of spacetime are, I believe, at the moment creating barriers for science and technology, instead of inviting us to explore the cosmos through an entirely new paradigm.

These, then, are the barriers that I seek to transcend in this book. I address the problem of conquering spacetime, from diverse genres such as biological perspectives (genes, DNA) to electromagnetic dynamics and physics (rolled-up dimensions, 4-D). I suggest that the only way human beings will be able to travel this mysterious cosmos is through holographic means, becoming appearances themselves to other cosmic beings. Delving deeply into biology, physics, and holographic dynamics has by no means exhausted my wonder about the mystery that underlies the cosmos. I have come to believe that while we can investigate our relationship with the cosmos through appearances and science, the mystery underlying the cosmos remains an ever-expanding one.

Although I wrote the book by myself, I was helped in the final stages by the advice of a dear friend, Bill Mathews, who, in reading my manuscript, no doubt was surprised by my rather unconventional ideas. He, along with my son, David McRobert, attended to the technicalities of this digital age to make sure those technicalities did not wear me out. I am also grateful for input from my son Charles McRobert. I thank them all for being there for me.

INTRODUCTION

In this book I philosophize from a multidimensional perspective, integrating philosophy with physics, biology, astrology, and popular material. I ask whether the appearances some people claim to see are projections solely of human imagination or whether these visions are part of a substantial cosmic force. I ask, too, if the two are interrelated dynamics. In other words, are appearances embedded in the hard drive of a biological cosmos that acts upon us, either consciously or unconsciously? And if so, is it the same system (or a facet of it) to which our founding prophets of religion, other seers, and even our scientists in the past responded—and today are responding?

Prominent scientists presently are working on deconstructing the physiology of the brain in order to pinpoint the ephemeral experience human beings have of being conscious. Some of them even believe that they will discover a physical substance that either underlies or constitutes consciousness. Along with David Bohm, I strongly suspect that the universe in which we are caught up is based on quantum principles and thus is holographic in nature and that the element in consciousness that appears to be missing (because we cannot account for it neurobiologically) is missing because it is quantum principled.

Intelligent people throughout the ages have reported encounters with many different orders of phenomena—they have had encounters with *beings* or have seen visions of other "worlds" that they cannot explain. Is this because human consciousness might be part of a substantive underlying holographic cosmic consciousness that acts upon us consciously and unconsciously (during dreams)? We do not know when appearances and visions began in Homo sapiens, but we do know that the Sumerians left a myriad of writings on the subjects of spells and potions to cure them; we know, too, that the Middle Eastern area, primarily India, relied on astrology to guide their daily lives and that they wrote about their historical encounters with "gods." Today, we either respond to phenomena or appearances by relating to them as forces within or without ourselves. True, not everyone believes in these phenomena, particularly those who are scientifically minded; still, we do stimulate our imaginations by concretizing—by making real—our visions, dreams, and appearances and thus collectively become creators of ever-changing worldviews.

I assume that visions and appearances occur to people and examine some of the literature in this respect in order to discern just what may be happening cosmically to produce them. When prophets, mystics, philosophers, theologians, physicists, and psychologists argue, through the framework of subjective/objective language, about how something impinges upon thought from within or how some force acts upon them from without, are they simply reporting to us how their neurovisionary projecting brains work? If this is the case, how do these reported dynamics, from these different disciplinary areas, stand up to the neurological evidence about how the brain perceives images and subsequently how thought works? I assume, with Antonio Damasio and many philosophers of bygone centuries, that without impressions, sensations, and emotions that produce images, concepts could not be formed. With this in mind, I examine consciousness and its tango with appearances but also with a view to explaining the phenomena as a projection from the cosmos.

Assuming that our experience of consciousness is part of a holistic cosmic conscious force, then there are a myriad of questions with which to tangle. I begin with the most obvious and popular ones, such as: Are UFO projections and other alien visitors from space riding on the beams of this greater cosmic consciousness? If so, what causes these projections from some other "world" to interfere with the visual system of particular individuals? Does it have something or nothing at all to do with the visual brain? Why do certain women and men report that they have been subject to alien rape? Is this physical sense of rape an indication that they are responding to the call of *birthing a cosmic intelligence*? Is it this system or another facet of it to which our biblical prophets responded? Most important, does a technological instinct rule the human race, and is it naturally propelling us toward the scientific quest of exploring the cosmos?

To delve further into this, I discuss issues and questions such as: Does a technological imperative rule the human race, as Heidegger believed? I ask whether a technological *instinct* (as distinct from Heidegger's *essence*) propels us toward discovering ever new scientific means of exploring the cosmos and if so, for what purpose or end? One of the examples I use involves award-winning inventors and how they produce holographic images that look as lifelike as the real objects, and I argue that if techno-artists can produce ephemeral, lifelike holograms, then why not the cosmos? As we become psychologically and physically involved with human-made three-dimensional environments and holographic images, I contend that we will be in a better position to understand the role of both psychic and cosmic appearances and why they are necessary phenomena for biological cosmic survival.

Further, if human beings are technologically close to producing holographic environments on earth, is it reasonable to assume that the transmitting of holographic environments and even the mind-boggling concept of teleportation could have been and still is the choice of aliens somewhere in the cosmos and used by them to transcend spacetime travel? If an appearance manifests itself by circumventing space and time, by interacting with genetic material through electromagnetic holographic transmissions, have we earthlings arrived at a point where we, too, are poised to transmit ourselves as holographic images to aliens elsewhere in the cosmos?

There are implicit messages running throughout the book that portend the urgency of changing our present-day ruling paradigms; for example: 1) that we must rethink our relationship with the cosmos which, for the moment, rules our conceptions of God, UFOs, and other types of appearances; 2) that we must understand the importance of appearances and why they are confirming that we, as they, are biologically part of the cosmos; 3) that we must begin to imagine spacetime travel in ways other than how we do today; 4) that we must learn to transmit holographic messages to aliens.

CHAPTER 1

Cosmos as a Psychic Control System

Since this book is about appearances, a good place to begin is with the very popular unidentified flying object and the abductee phenomena. What are these manifestations, these appearances? Who or what is responsible? What is the controlling influence involved in these similar but different visions, appearances, and manifestations that have appeared to people throughout the ages? Is there a way to explain the cause and the effect seen? We will come upon these questions and others, time and time again in the book.

I propose to take a twenty-first-century tack in my approach to this subject by asking whether these manifestations are stored in an individual's DNA and therefore are projected by the individual himself, or whether this is natural phenomena that occurs because of cosmic electromagnetic interference and therefore is projected at us. Does the projection of appearances work both ways? Are we, therefore, capable of tuning into other spacetime dimensions within ourselves or into spacetime cosmic electromagnetic interferences? Are we also capable of visually projecting the holograms we see by superimposing them onto our naturally seen three-dimensional environment? More succinctly, is our biology simply part of a cosmic biological stream?

As we address these many questions be prepared for a journey both factual and speculative, mythical and scientific, and laced with just enough philosophy to suit a digital age.

Vallée: Dimensionality and Spiritual Symbols

A myriad of accounts of unnatural or phenomenal sightings are easily found on the Internet; I will not attempt to summarize them here. Rather, we will begin our analysis of what appearances may or may not be with Jacques Vallée, who points out in his book *Dimensions*[2] that UFO-type of experiences have prevailed throughout recorded history and that the similarity of the experiences of little men (or people), objects in the sky, and beams of light that transport humans to other realities ought to no longer be ignored as simply figments of the imagination. French born, he studied astrophysics before receiving a doctorate in computer science in the United States and considers himself an information scientist, not a physical one, although as one reflects upon his work, one deduces that he actually has combined the two. He has devoted many years to studying phenomenon that appear in the skies and believes that these manifestations of UFOs and aliens are not extraterrestrial but rather other-dimensional, beyond space and time as we know it. He states:

> The UFOs are physical manifestations that simply cannot be understood apart from their psychic and symbolic reality. What we see here is not an alien invasion. *It is a spiritual system that acts on humans and uses humans.*[3]

It is, he believes, part of a psychic progress that we have not yet mastered and do not understand that "an alien form of intelligence of incredible complexity is communicating with us *symbolically*."[4]

Vallée's is a coherent statement about what the UFO phenomenon may be—is it a spiritual system or, as he prefers to refer to it, a "control system"[5] acting upon us? Over the years he has had many insights into the reasons for why UFOs might manifest themselves. These reasons go beyond the claims of UFOlogists with whom he once sided when he began his writings on the subject. As already noted, his insights into this mysterious phenomenon now extend into what he suspects could be breakthroughs from a multidimensional universe, a phenomenon that is paranormal in nature and is manifesting itself *for a reason* in our modern technological age.[6] Essentially, he believes that these phenomena are happening in order that we create a new order of social reality. According to Vallée, our future worldviews will be shaped by these manifestations once we break through their patterns and decode their messages. He states in an interview that although we do not know much about electromagnetism, microwave radiation, or colored pulsating lights and/or their other effects on the brain, any of these could be affecting the consciousness of people who see such phenomenon.[7] He makes an important point with these speculations, some of which we will follow up on later in the book. We should note that Vallée defines consciousness *"as a process by which informational associations are retrieved and traversed. The illusion of time and space would be merely a side effect of consciousness as it traverses associations."*[8] The idea of an alien technology permeating our world in some sophisticated manner that we do not yet understand in order to control, contain, or condition us has occurred not only to Vallée but to others as well who independently study the phenomenon from quite different perspectives.

Let us examine what Vallée has to say in *Dimensions* about existent mythologies that expound either encounters with little men or with tall, luminous, unearthly men and their flying machines, which turn out to be astonishingly similar in description throughout the ages, although adaptive in technological imagery to their cultural time.[9] Vallée quotes Edwin S. Hartland, who wrote the following in 1891: "Man's imagination, like every known power, works by fixed laws."[10] Indeed, it is this search for a pattern and for the laws of the imagination that Vallée attempts to map, advising those who study UFO phenomenon to do the same by building on the work of foregoing scholarship on the subject. Although he has sorted through much available historical data, he claims there is still much more information that needs to be compiled in order to break through the esotericism of the symbolism that confronts us. He sets out his question this way:

> The faint suspicion of a giant mystery, much larger than our current preoccupation with life on other planets, much deeper than mere reports of zigzag-ging lights—perhaps we should try to understand what these tales, these myths, these legends are doing to us. What images are they designed to convey? What hidden needs are they fulfilling? Are there precedents in history? Could imagina-tion be a stronger force to shape the actions of men, than its expression in dogma, in political structures, in established churches, in armies? If so, could this force be used? Is it being used? Is there a science of deception at work here on a grand scale, or could the human mind generate its own phantoms, in a formidable, collective crea-tion mythology?[11]

Vallée's careful, long-term research on the subject of UFOs has brought him to the conclusion that what we are encountering are not extraterrestrials from other planets but aliens from beyond space and time—from another dimension. Materializing in our reality from another reality, they are, nonetheless, physical in nature,[12] inviting us to transcend into this other order of superior reality, even though they proceed in absurd and bizarre ways. The messages we receive from what could be a multidimensional universe, according to Vallée, are symbolic in nature. Whether they are good or evil, he does not speculate, although he appears to find them positive symbols that may take centuries to decipher because our brains have not evolved enough for us to understand these symbols.

Vallée provides us with many examples of reported UFO experiences. Consider the one he cites of individuals who see UFOs in a field in the country. Standing outside and around the UFOs are two to three men, sometimes as many as five, who are dressed as spacemen. Their suits are shiny (in what today we might refer to as metallic) and are sometimes described as inflated; the spacemen wear helmets and/or backpacks/frontpacks. (The Mayas have carved stelae of what look like spacemen dressed in this manner—one sees them in Mexican museums—and there are a smattering of other artifacts of this nature that exist throughout the world.) From Vallée's examples one could surmise that the psyche is projecting into the future or accepting from the future a destiny that awaits humankind. Or the psyche could be recalling what I would identify as stored, genetically based appearances from the past (something that we will discuss in a later chapter). Vallée suggests that the UFOs encountered appear to be a product of the limits of the imagination of the epoch in which it is experienced. None of the UFOs seen in the early centuries, he tells us, would have been capable of flying. They are manifestations of machines that physically appear but are limited to the technological imagination of the populace at that particular point in time. For example, in the twelfth and thirteenth centuries these vehicles appeared boat-like and often dangled anchor-like objects that got caught in church steeples.

Symbolic Conditioning

Vallée believes these other-dimensional aliens are trying to slowly condition us by plying us with specific information through their control system. It is a conditioning that is meant to affect human beings socially, so in a sense, his project is also about a morally evolving consciousness. As already alluded to, he warns that one ought not to confuse an abductee's experience of aliens with extraterrestrial beings from some other planet, galaxy, or parallel world related to space and time. This is not about leaving the body behind and becoming an abstract being. Conflating these two quite different genres of experiencing aliens keeps us from addressing the true mysteries that underlie the ruling orders of our spiritual consciousnesses. Vallée's theory suggests that there is a common denominator between sender and receiver, and so we are not left with a Descartian schematism—an incomplete version of an alien dimension out there in some unknown beyond and a human receptor down here on earth. In Judaism and Christianity, for example, we have grace filling this in-between gap between a God beyond and humankind on earth. In Vallée there is this possibility that electromagnetism, pulsating lights, or microwave radiations could be filling the gap. By going beyond this schism and accepting these appearances as possibilities, whether they are part of the abductees' experience or someone else's version of extraterrestrial beings, I will attempt to establish a biological paradigm, one that borrows but differs from both physicists' perspectives of multidimensional universes and from Vallée's idea of an information or control system.

Psychic Projection of Physicality

Since we have been referring to the abductee phenomenon, we need to include some examples of their experiences and for this I turn to John Mack's work on the subject.[13] Mack, a Harvard psychiatrist, interprets what he refers to as abductees' self-transcendent experiences in a positive light, even though he warns that we should not ignore the fact that abductees sometimes appear to suffer from the Stockholm syndrome. Under hypnosis, they have recounted that after feeling great aversion for their abductors, they develop a great love for them; that is, abductees seem, in time, to identify with their torturers ("torturers" because they extract biological samples from the abductees in rather painful ways) and end up thinking of them as good and benevolent. The main purpose of their abductions is hinged on creating a hybrid race. Mack also reports that many of his patients are converted into becoming spokespeople for these alien abductors, who instill in the abductees the desire to save the earth's ecological systems; to save our planet while there is time to do so. There are reports of a spiritual awakening, of a marriage between human and alien that is not unlike that of the human/divine model in Orthodox Christianity.

It is also possible (and he has been accused of this) that Mack projects his own interpretation of the psychological dynamics experienced onto the abductees in order to help them come to grips with their proclaimed experiences. According to criticizers, this can happen during hypnosis or in conversations with patients outside of hypnotic sessions. He might implant the suggestion that something creative and good can result from all the terror, disgust, violation, rape, etc., that the abductees experience. Mack believes that the messages the aliens have instilled in the abductees, through the programming of their psyches, has everything to do with introducing to earth a new hybrid creature in order that human consciousness on earth be expanded to embrace a cosmic one. "The earth would become the jewel in the crown of our being, the place where we experience once again our connection with a cosmic Source from which we have become too separate."[14]

If Mack is right, then abductees could be at the forefront of accepting the fact that a hybrid being can be produced and the earth's ecological systems saved, so that earth could someday harbor a consciousness that would be an inspiration to other be-

ings in the cosmos. As Mack argues, these "chosen" human beings recount to us that their otherworldly experiences have lead
them to believe that this kind of hybrid procreation is their duty.
That is, if life is to continue on earth, or elsewhere on another
planet, or in another dimension beyond our concept of space and
time, it will lie in the superior spirituality or intelligence of this
newly created hybrid being. We know, of course, from our ongo-
ing virgin-and-child stories, that this kind of thinking has occurred in history before—and that it is among our oldest, most
fundamental myths.

So Mack's theory of ecological saviorism[15] in some aspects
mirrors Vallée's theory that the entire UFO phenomenon is based
on the absurd, on dynamics that we do not yet understand *but
which are there for a purpose.*[16] All we can do is accept these recounted experiences of appearances and try to discern an underlying logic to the mythological circuitry with which they continually
confront us in each epoch. At best, both men believe that what
we are encountering is some sort of a dimensional breakthrough.
Vallée thinks of it as a metalogic that is beyond the capacity of
the human brain's understanding at present. As he puts it, mythology rules on a social level over which politics and scientific action
have no power.[17] Mack, on the other hand, stays largely within the
parameters of a psychologically evolving consciousness, and his
concluding remarks refer to the abductees' experiences as "a kind
of cultural ego death"[18] that prepares them for the next step in their
conscious evolution.

So whose conditioning system is this, we might ask? Is it like
a Platonic idea that has always been out there, waiting to come to
maturity, and if so, is it a good idea or a bad idea? Or is it cosmically inspired, and if so, must we assume that all cosmic ideas are
lawful and so necessarily good? Of course, we ought not to ignore
that cosmic ideas could be seductively fractal, chaotic, and lead to
cosmic annihilation.

Jung, Archetypes, and Archetypal Projections

Michael Heim has argued that the gray-men some people claim to see are our own projections from our future, a fascinating idea from which I have benefited greatly. Heim's idea is distinct from Jung's idea that UFOs are material projections of the psyche. As Heim writes in his book *Virtual Realism*,[19] perhaps what we see in the psychic processes underway is that the two experiences—the archetypal archaic and the archetypal future—are very closely related, with the future being more difficult to envisage. Jung, Heim professes, would have had it right, had he not neglected the future.[20] Aniela Jaffé writes: "Jung has explained the UFOs as a projection of a psychic content (of wholeness) that has at all times been symbolized by the circle. In other words, this visionary rumour, also present in dreams of our time, is an attempt by the unconscious collective psyche to heal the split in our apocalyptic age by means of the symbol of the circle."[21]

But there is another side to the claim that we are human hybrids created by extraterrestrials or aliens from other dimensions that could also be considered plausible. Are we being prepared, psychically and physically, by these cosmic consciousness-interfering projections from the future to leave this planet and venture into the cosmos? Funnily enough, Jung speculated on these aspects too. He recounts the following dream about UFOs as a dream typical of UFO accounts of sightings:

> In one dream, which I had in October 1958, I caught sight from my house of two lens-shaped metallically gleaming disks, which hurtled in a narrow arc over the house and down to the lake. They were two UFOs (Unidentified Flying Objects). Then another body came flying directly toward me. It was a perfectly circular lens, like the objective of a telescope.

> At a distance of four or five hundred yards it stood
> still for a moment, and then flew off. Immediately
> afterward, another came speeding through the
> air: a lens with a metallic extension which led to a
> box—a magic lantern. At a distance of sixty or sev-
> enty yards it stood still in the air, pointing straight
> at me. I awoke with a feeling of astonishment. Still
> half in the dream, the thought passed through my
> head: "We always think that the UFOs are projec-
> tions of ours. Now it turns out that we are their
> projections. I am projected by the magic lantern as
> C. G. Jung. But who manipulates the apparatus?"[22]

Recounting this dream was Jung's way (he states) of describing the reversal of the relationship between ego-consciousness and
the unconscious. A reversal of this nature is meant to "represent
the unconscious as the generator of the empirical personality. This
reversal suggests that in the opinion of the other side, our unconscious existence is the real one and our conscious world a kind
of illusion, an apparent reality constructed for a specific purpose,
like a dream which seems a reality as long as we are in it."[23] In
reflectively speculating about his dream, Jung wonders whether it
was really something in itself, or whether there was someone controlling the projections of the reality in which we live. In the end,
Jung prefers to postulate that the dream was nothing more than an
illusion and that the unconscious was attempting to tell him this, in
order that he could tell us.

There are psychiatrists/psychologists who believe that abduction experiences are really a form of a dream (i.e., similar to
a lucid dream). Some rebuff abductees' reports, saying that their
memory is being manipulated by the psychotherapist who has
hypnotized them, while still others insist that these hallucinations are created by disturbances in and of the brain. Most in-
vestigators of the phenomena, however, remain stymied by the

universality of the reports and because of it, tend to bracket the experience into a Jungian realm of a collective unconsciousness, speculating that it has resurfaced in the collective consciousness because of exposure to modern cultural ET films and/or films dealing with the supernatural and so on.

But can we assume that these experiences are dreams made out of the common archetypal stuff of dreams? True, the appearances appear to be *somewhat* like dreams, possibly (*and this is my thesis*) because both kinds of dreams are produced or interfered with by the electromagnetic frequencies of a cosmic stream of consciousness. As in archetypal dreams, there is a message that is relayed through a theme, which in the abductees' case is their experience with the aliens. The messages they receive are not verbal ones; rather, mind-to-mind messages are reported by abductees, or the messages are mimed. The mind-to-mind communications used by aliens in these situations to inform abductees of the future can be interpreted as being like dreams, self-contained, and part of the same cosmic communication system that is used by the brain to project archetypes of the past and/or present. How this happens in dreams and whether these images are made visible to us through holographic projections is subject to further consideration later on.

Most of us already have a pretty good idea of what archetypal dreams are all about, but few of us remember Jung's adage that the *archetype is a portrait of the instinct*, and that hence the origins of archetypes are to be found in our flesh and blood. If we are to understand the nature of the appearances with which we are dealing, we need to delve further into this archetypal language. One way to address archetypes is to do so from the point of view that instincts from the future are part of our genetic inheritance, that they have always been part of our human makeup since at least the time of mitochondrial Eve, the woman out of Africa who is supposed to have been our first ancestor some 200,000–450,000 years ago[24]. This, for example, would be part of Christopher Wills' theory in his book *The Runaway Brain*[25]—that instincts for our human potential are there from the beginning, in the very first cell from which we evolved, but they just get shuffled around in various combinations and permutations. Although I do not agree with him that the human being developed entirely through evolution, as Darwin would have it (Wills' is a neo-Darwinian view), I do agree with Wills that all of our instincts for humanness are programmed in the genetic data that constitutes who we are today, and *that would mean any "divine" or "alien instincts" that we now possess in our genotype,* something that we will discuss in later chapters.

Symbolic Archetypal Language, Past and Present

There are certain aspects about the spiritual or intellectual control system that we are examining that are curiously unsophisticated. It is interesting to note that whether symbolic language comes to the psyche from the past or from the future, it is very much rooted in overarching notions of sexuality and reproduction. In many ways, Freud was right: everything can, indeed, be reduced to preoccupations with the sexual—intercourse, conception, pregnancy, hybrid creations (human/divine). Only recently have scholars in the fields of psychiatry or psychology given the nod to instinctual images that favor *bodily* biological manifestations rather than solely supporting images produced only by the brain.

At this point, because my claim is that appearances are rooted in our cosmic biology, which is the central thesis of this book, we need to examine a few of the foundational Jungian concepts vis-à-vis the machinations of the psyche through archetypal images. Jung's approach keeps appearances both physically and psychically grounded—all of our projections are part and parcel of the mind. But although Jung concentrates on the psyche, his is not a one-sided view. There is a duality in play in Jungian dynamics, and we must take our clue from it, because if the archetypal dynamics occur as projections from the psyche into another dimensional spacetime, we can throw a different light on the abductees' experiences and also on appearances in general. The psychic energy or libido, Jung describes, operates within a hypothetical psychic space. Within its structure a dynamic polarity, based on opposing tendencies, oscillates. For example, Jung would probably see the abductees' love-hate relationship with the aliens as a polar-based intra-psychic relationship. The energy created by this polarity helps a person experience a natural drive toward wholeness. But because psychic energy is not a pure concept, it is generally suffused with physical forces—we experience it in sexual, vital, mental or moral form.[26] Thus for Jung, the psyche is very real. It is as real as the physical, having its own structure and subject to its own laws.[27] Within the psyche, the spiritual energies that propel a person toward individuation tend toward the conceptual but can never be purely conceptual.

Another aspect we need to consider further, if we are to better understand the abductees' experiences, is Jung's claim that the *archetype is a portrait of the instinct.* For example, when we dream, the instinct's formal aspect takes on archetypal shape that is, generally speaking, a human shape. Archetypes are difficult to grasp, Jung points out, because the human mind never invented them—*it simply inherited them.*[28] Archetypes, according to Jung, contain all of human experience, right back to the remotest beginning of time. These remote unknown beginnings carry within them the source of the instinct, which expresses itself by assuming its opposite form, the archetype.[29] Archetype is simply "the instinct's perception of itself"—a self-portrait of the instinct.[30] It is the archetype that makes it possible to translate physical factors into psychic factors. Jung did not believe that aliens were real but rather that the instinct-archetype materialized them in order that they be transformed into spiritual factors. (I might point out that Jung's idea of spiritual and Vallée's idea are not on the same wavelength. Jung's idea of spiritual emanates from within a biologically bound psyche, while Vallée's moral but abstract spirit remains bound to informational space.)

Symbols are what we become aware of in consciousness as images, whereas the archetypes running the show contain the numinosity that allows for, through their empty form, the creation of symbols. Symbols are the manifestations of the excesses of spiritual energy found in the psyche. When Jung looks at the archetype, he is looking at the mental mechanism itself; when he looks at the symbol, he means it to be an external expression of an empty form—something that is incarnated.[31] So far as Jung is concerned, the ineffable quality of the archetype, represented by the symbol, speaks only in the language of metaphor, in myth, in parable, in creations of art, in music, in visions, and in dreams. In his view, the symbols produced by the instinct-archetype revitalize or energize personal growth. This occurs because the symbol unites opposites, something the intellect cannot bridge by itself. Applying this logic to the abductees' experience, the UFO becomes the symbol produced by the psyche in order to unite, to bridge the gap between human beings and aliens. In effect, what the abductees see is a symbolic-but-materialized UFO from which aliens appear. This is the psyche's way of incarnating spiritual energy—a way that plays a major role in establishing a larger base for consciousness and personality.[32]

Interestingly, the dynamics described by abductees fit into these Jungian categories quite easily. For example, the abovementioned tension of the love/hate relationship between the abductee and the alien abductor is a relationship that is eventually reconciled in most abductees' minds as being a positive one. Dynamics seen as spiritually absurd by Vallée in his book *Dimensions* could be said to be part of a normative yes/no Jungian dialectic through which the psyche seeks self-transcendence. If there is a tendency in the psyche to oscillate between the familiar archaic and teleological goals, then surely the abductees' experience falls neatly and unquestionably into this category.

Divine Births; Hybrid Creations

Not only the abductees but we, too, as a historical culture, are very much preoccupied with hybrid images. From the ancient past we have Sumerian accounts of hybrid creation. Almost identical myths have been unearthed most recently in Mayan lore. We are inundated by ancient great-mother statues—goddess women pregnant with child and cultural myths of virgins who conceive by the grace of god(s). Today, we can artificially inseminate virgins and/ or grandmothers, but our fixations are still with the mythological, with sex and conception, with the zygote and its divine genetic inheritance, to say nothing of the idea that the Bible implants in us that we are made in the likeness of God. It is not because of pure biology that the fight of the right-to-life movement goes on in the United States and elsewhere in the world. The right-to-life movement feels a strong connection to the divine inheritance of the zygote—and of course, they are right in this respect; there is something sacred about our DNA and not only at the moment that it becomes part of a zygote.

Surely all this conscious and unconscious preoccupation with the hybridness of our "test-tube creation" myths must be saying something to us symbolically about our so-called spiritual natures—something to the effect that they are very much a real part of us, inside of us, and not just outside of us. Vallée, unlike Jung, thinks that this outside aspect of spirituality is controlled by a system of laws that possibly originated in another dimension, which breaks through into our own three-dimensional one to control us. Here, I find him very Ricoeurian in *context* but not in *content* and still, by and large, within our Western tradition's belief system that something unknown acts upon us. Like Vallée, Paul Ricoeur argues, in his seminal book, *The Symbolism of Evil,*[33] that the primitive psyche experiences otherness through feelings of dread, feelings of guilt and of defilement (the stain of sin). This otherness acts upon the collective as well as the individual primitive psyche. It descends upon us from some other dimension, some beyond, and can be profoundly felt by a person or community, even though its origin is not understood. These primitive experiences are the source of our myths, symbols, and religions, according to Ricoeur. To a great degree, one has the feeling that Vallée, however unknowingly, is also caught up in this kind of intellectual myth, which remains well outside of anything biological, let alone bound to a cosmic dynamic.

Playing a great role even today, there is a school of thought that continues to convince most intellectuals that UFOs are mythical constructs of the mind. Rightly identified as otherworldly projections of the mind, these appearances are wrongly interpreted as superstitious nonsense. Vallée believes that the UFO symbol is beyond our lived three-dimensionality and that it is this other dimensionality that is very pertinent and important to decipher in the UFO experience. If we put aside the old ideas of religiously inspired psychic projections for new ones that are not religious and not purely physical theory, and we reconsider them from our present knowledge base, we have to ask how the breakthroughs by UFOs into our spacetime dimension occur. Are they breakthroughs from the cosmos or breakouts from the psyche? Are they instinctively provoked by electromagnetism? What in our genome allows us to accept and project these images as though they were *material* onto our reality? If we believe that we are continuing to evolve intellectually, imaginatively, and not necessarily only physically, then we ought to welcome these material manifestations of UFOs, because they fit into the evolutionary scheme of things. One form of this imaginative evolutionary schematic that can be specifically identified as influenced by these UFO manifestations is our push to investigate our planetary system, with hopes of further exploring our galaxy.

The UFO phenomena seem to be right on target in proffering an evolutionary image of modern cosmic content. With the scientific mind in charge of the cosmic future adventure, this is, indeed, an appropriate time for these other dimensional breakthroughs to flood the skies above and the minds below. It is not surprising that the language of instinctual images has begun to proffer representational symbols of alien beings that are no longer luminously god-like and fashioned in our own image or even as part animal, as in Mayan and Egyptian glyphs that, by the way, give us an enormous clue that points to the transgenic stuff we are made of. The early biblical reports that describe the form of angels and other messengers from the gods or the fairytale reports of little people that span the mythology of the world's cultures are quite different from the symbolic forms that abductees encounter. That a new form is manifested by the psyche—or to it—is evident in the consistent descriptions today by abductees of gray aliens with their meager bodies; big, brainy heads with large dark eyes; nostril-dominant, pushed-in noses; and slits for mouths. Are our brains really responsible for these forms and the projectors of them? If, as we have seen above, an abductee's brain processes can retrieve the old but also create new representational forms from the instinctual images available to it, it is more than likely that the abductee's description of a new form of hybrid being is the only one that she is capable of conjuring up, which turns out to be quite limited and abstract.

Again, the fact that these aliens from the future are somewhat repulsive to look at invites yet another possibility as to why abductees seem preoccupied, in Mack's opinion, with preserving what is left of the earth's ecological systems. It must be disturbing for abductees to see what is destined to be left of their human race in this unattractive form. It is as though their encounter with the alien hybrid leaves them with a longing to preserve the human/divine form in all its glory and as they have always known it. That is my guess as to why the issues of sex, hybrid reproduction, and ecology are so ineluctably bound in these accounts. Thus, if there is a spiritual control system in place that issues from another dimension, and if this other dimension has been programmed into our instinctual imagery from the start so that we might someday be capable of identifying it, then it is arguable that some of us, such as the abductees, are reminding the rest of us that this is what a future hybrid form could look like. If we do not like what the abductees see then we ought to do our best to cherish the human/divine form that we have, choose another path through the hypertext of choices available to us now and create a different cosmic story.

This, however, may not necessarily be the reason for the abductees' prophetic-like desires to preach ecological consciousness to us. The ecological emphasis might be due to the fact that an abductee's consciousness cannot yet reconcile the *idea* of a new hybrid form with an *image* of it—the alien's unattractiveness. If these aliens are from the future, then it is not surprising that abductees describe them as they do. The pre-programmed instinctual image available to them that represents the hybrid form comes from the abstract recesses of the human mind. In their minds it may be perceived as an alien form that is gray and intelligent or, as some abductees describe it, as an image of pure light. The image of form the mind proffers is an abstract one, the result more geometrical. Consider the triangular face for an alien and stick-like form for its body. In this sense we could say that there is a psychic interplay between archaic and teleological tensions in the genome that results in an incomplete concept being visualized by the mind, just as Jung claimed above. How the mind creates these visualizations is still being studied and debated, but of course, the starting point is always biological.

Kantian Dynamics

It is easy enough to slough off the appearances discussed above as figments of the imagination without considering biological factors, but before we do that, consider the fact that philosophers, such as the eighteenth-century David Hume, were already arguing that ideas are reducible to sensation, while others, such as Immanuel Kant, who lived a little later in the same century, pondered abstractly about the formation of appearances and their relationship to sensation. In *The Critique of Pure Reason*,[34] Kant posits that we are given objects or their appearances through the sensibility. These appearances are transmitted through the intuitions, which in turn are processed through the understanding, which in turn gives rise to concepts. Kant states:

> That in the appearance which corresponds to sensation I term its *matter*; but that which so determines the manifold of appearance that it allows of being ordered in certain relations, I term the *form* of appearance. That in which alone the sensations can be posited and ordered in a certain form, cannot itself be sensation; and therefore, while the matter of all appearance is given to us *a posteriori* only, its form must lie ready for the sensations *a priori* in the mind, and so must allow of being considered apart from all sensation.[35]

Now this is difficult-sounding, convoluted language to our twenty-first-century sensibilities, so without delving into Kant's critique of reason and/or a critique of Kant's critique of reason here, we will just consider how Kant's philosophy describes appearances that present themselves to the mind, keeping in mind that Kant holds the thesis that we cannot know the thing-in-itself, just its appearance. (In chapter 3 we will deal more directly with Kant's views on the phenomenon of 3-D or projected appearances.) Even though Kant knew nothing about DNA and the genome, he certainly zones in and scores on the fact that innate to the human being's mind are the ever-transient forms for appearances. They are always there *a priori*, but they need to be filled in materially through experience. Accordingly, it is arguable that this is precisely what the minds of the abductees do—they furnish us with a visual picture of how their minds work, from instincts of the possibility for to a projected appearance of a form that lies within them.

As though to provide us with a screen upon which appearances need to appear, Kant reminds us that without *a priori* intuitions of space and time (conforming to outer and inner intuitions), our minds would see no objects or appearances. Interestingly, when abductees try to explain where they have been during the time when others notice their absence, they usually do not relate to the loss of time. Does this mean that their intuitions for time and space have somehow been tampered with? In folklore, the loss of time is often reported in years. The abducted person generally thinks he has only been gone a short period of time. Vallée points out that in folkloric tales there is sensitivity to the relativity of time, which, from the examples provided by Vallée, would precede Einstein's discovery of it.

What is constant in most abductees' stories is the report of being in another dimension, where space and time are collapsed. Or they speak about aliens breaking into our dimension from another one. When we apply Kant's logic to this—that time and space are subjective, intuitive experiences—then we must take into account that abductees, although they may appear to be beyond our intuitions of space and time, are simply in another intuitive or instinctive dimension of it. It is as though the abduction experience is one that unfolds for the abductees through their bodies' cellular structure; more precisely that they are in another dimension of space and time that is also encoded in their DNA and that this is what is being projected within or without them by their minds. Is this why the abductees feel as though the time and space they know are collapsed in this alien dimension? In biological terms, can it be that it is simply collapsed temporarily within their instincts, while they have transcended into another instinctive dimension of it?

But let us focus once more on the appearance itself. Just how does it occur? Similarly to Jung, Kant addresses a one-way projection that originates from within the person. We must raise the question of how these appearances can take on materiality if there is no other activating or initiating process underway. How is our visual brain, whether asleep or awake, involved? What motivates the phenomenon to break out of our minds, and is the phenomenon instinctively activated by the mind to do so? Is there a broader reciprocal, cooperative interaction underway here? Can the entire dynamic be understood only if one situates it in a cosmic consciousness; that is, if we understand our DNA to be part of the cosmos and immersed in its energies? Again, we are ahead of ourselves, but these are questions that will be answered as we proceed to future chapters.

To be sure, Kant is on to something with his position that the dynamics of sensibility are unable to proceed beyond the innately programmed instinctive forms without bumping into reason's ideals and thus into an emptiness of forms or a distortion of materiality. He argues that as we exhaust our sensibility's capacities to account for appearances, we run into reason and cannot provide materiality for these abstract forms. Hence, Kant's famous phrase: that ideas are really empty forms unless they become images. As Jung proffers, so does Kant—a brick wall is hit unless our instincts can provide the needed images for the innate forms. Recall that Kant's general theory of appearances that emanates out of our intuitive space has to do with his belief that we are pre-programmed with these forms and thus, when they appear in or to our minds we easily fill in their matter against intuitively created space, the way children do the forms in a coloring book. However, he also believes that our imaginations cannot go beyond the instincts with which we have been programmed. We are instinctively limited. All this makes a great deal of sense in light of the UFO experience, and it is particularly relevant in respect to the abductees' description of the aliens' physical body.

Pushed to a natural instinctual end, we are left—or so the abductees' description of aliens implies—with an image of an intellectualized bodily form, free of all its instinctual, natural, physical beauty that makes humanness what it is. Indubitably, we ought to pay attention to the stories that prevail that tell us we are created in the image of God, for it would appear that we are, indeed, pre-programmed to contain all of the instincts natural to our own species but also to those of the "divine" species that created us. If it is our turn to pass along intelligent life in the universe, we are charged with the dilemma of imagining a hybrid form that we might create but for which we have no existent instinctual image—at least in Kant's view, except, of course, the one the abductees leave us, which is not very savory. Perhaps this is the reason why subconsciously, abductees opt in favor of stressing both the creation of a hybrid child and the ecological angle of saving the earth. All this is part of the absurd oppositional spiritual logic of the abductee experience that conforms in dynamics with the intrapsychic Jungian psychology discussed above. My own suspicion is that falling back into familiar imaginary, like the earth's environment, is a distress or default ploy of the mind; an attempt to fold itself back into its usual realm of sensibilities away from the unpalatable images of aliens that present themselves to it.

Scientists, however, push forward beyond this default setting with their vision of transcending this earthly environment and finding a suitable new planet elsewhere in the cosmos, where humankind—inhabiting a suitable new body, no doubt—could migrate someday, when necessary. Although, perhaps unbeknownst to them, they seem to be gravitating to a black-hole dimension that is otherworldly in its character.[36] Contrary to Kant's belief, perhaps his idea of pre-programmed instincts—the abstract forms that abductees try to fill in—can be filled in more splendidly. If so, then, there is no reason why human imagination could not also visualize the creation of a new hybrid form of life as easily as they now create robotic "life." Ideally, it would be an alien life form that they have genetically manipulated to abound with genes for intelligence or, as the abductees would have it, a higher order of consciousness.

CHAPTER 2

HOLOGRAPHIC DYNAMICS: POETS, MYSTICS, PHILOSOPHERS, AND PHYSICISTS

Appearances and/or phenomena are enticing subject matter for philosophers, mystics, and poets who have always wondered about the world in which they find themselves, why they exist, and what happens to their souls after death. This wondering has resulted in much speculation about otherworldly existence, which in turn has produced extraordinary music, art, literature, and poetry that describes such yearning. More recently, even string physicists have strayed into yet another version of otherworldly dimensions, something we delve into in chapter 8.

Consider that Goethe's poetic inspirations did not originate in a vacuum; rather, they resulted from his own unusual experiences. For example, he reports what could be construed as a sighting of a UFO or UFO lights (in the sixth book of his autobiography), not unlike what many people report seeing over cities in recent years.[37] He relates that while traveling as a young man between Frankfurt and the University of Leipzig, he saw the following as he walked up a hill behind the coach during the journey:

> All at once, in a ravine on the right-hand side of the way, I saw a sort of amphitheatre, wonderfully illuminated. In a funnel-shaped space there were innumerable little lights gleaming, ranged step-fashion over one another; and they shone so brilliantly that the eye was dazzled. But what still more confused the sight was that they did not keep still, but jumped about here and there, as well downwards from above as vice versa, and in every direction. The greater part of them, however, remained stationary, and beamed on. It was only with the greatest reluctance that I suffered myself to be called away from the spectacle, which I could have wished to examine more closely....Now whether this was a pandemonium of will-o'-the-wisps, or a company of luminous creatures I will not decide.[38]

After a sighting of this nature, it is not surprising that Goethe was inspired by light and eventually discovered an alternate theory to the dynamics of color by reinterpreting Newton's theory of light. Goethe, by the way, considered his work on light to be the most important contribution of his life's work.[39]

Phenomena can be addressed from many different philosophical perspectives, and here we will consider the eighteenth-century contemporaries Kant and Swendenborg. As we will see, Swedenborg, in his own way, describes the same genre of holographic dynamics that could be part of the imagination of any epoch. His visions still appeal to people from diverse fields of studies who are followers of the afterlife beliefs that Swedenborg posited. From our perspective in this book, we are interested in holographic images seen from any viewpoint in any era, particularly now, as technology has evolved techniques that produce holographic images that can be artificially projected and merged with our environment. We will examine some of these inventions in chapter 5 in an effort to throw light on the subject of biological holographic projection.

Kant and Phenomenon

For the moment, we will concentrate on Kant's reaction to Swedenborg's *Arcana Coelestica,* for it is widely held that Kant was drawn to, if not inspired by, Emmanuel Swedenborg's writings about otherworldly spiritual experiences.[40] While he did not attempt to disprove the validity of Swedenborg's philosophy, Kant wanted to accommodate both the natural and the intellectual world through reason alone. Kant did not agree with Swedenborg's concept of two separate but corresponding worlds (called "correspondences" by Swedenborg), a natural one and a spiritual one, with spirits crossing over into the natural one and communicating with chosen human beings (such as Swedenborg). What Kant agreed with was Swedenborg's concept of two sensibilities, interiorly and exteriorly experienced. Kant was determined to distinguish what he considered to be a this-worldly appearance, belonging only to the sensuous realm and life on earth, from the thing-in-itself (a noumenon), which could not be known. The best to which a human being could aspire was for reason to transcend his ego. Kant's transcendental ego belonged to an intellectual dimension with spiritual overtones, but it was strictly limited to this world.

Kant's inaugural dissertation, presented in 1770, dealt with these two worlds. In fact, it has been said that Kant plagiarized Swedenborg's ideas in respect to the dynamics of sensible intuition, as Swedenborg had written about this subject in his *Arcana Coelestica* as early as 1753. "The true and sufficient evidence of Swedenborg's influence on Kant is unmistakably shown in his 'Inaugural Dissertation on the Two Worlds' written in 1770, the year following the publication of Swedenborg's *De Commercio,* and in subsequent lectures on metaphysics and psychology which have recently been edited by Du Prel and Heinze."[41] In fact, Kant used the same terminology as Swedenborg—for example, *mundus sensibilis* and *mundus intelligibilis*—although other contemporaries such as Leibniz were writing along the same vein and using the same terminology which, to be fair to Kant, occurs in most epochs (for example, consider Wallace and Darwin; Leibnitz and Newton.)

Swedenborg wrote about "knowledges," vessels within man that could receive the rational things from "God with him."[42] That is, rational truths result from the inflowing of the divine, a divine element that is innate to a man but remains outside of him unless his heart understands it and appropriates it as divine. Swedenborg stated in the *Arcana*, "the things that then take place in the rational appear in the natural plane as an image of many things together in a mirror,"[43] while similarly, Kant later stated that appearances in space are the result of reflection,[44] so both men believed that appearances and phenomena were the result of man's innate ability to project reflective images from an interior source into a space exterior to himself. This is all rather abstract unless one is familiar with the work of these eighteenth-century philosophers. Familiar or not, as we proceed, their approaches to phenomena will provide us with evidence that appearances are intimately involved with our visual systems—dynamics that are important to our argument.

Kant begins his critique of pure reason by describing an innate notion of space, which can be thought of as a blank movie screen onto which vision projects phenomenon, somewhat like David Michael Levin (see chapter 3) produced his visions against a purposefully created black space. Kant appears to build on his early preoccupations with Swedenborg's thought. In the critique, Kant argues that the concept or notion of space is external to us even, though it is part of an innate (*a priori*) endowment. In order that phenomena manifest themselves formally to us as objects, we must project them onto this "screen" of empty space that emanates from within ourselves and manifests as external to us. Once the forms of sensible intuition are projected onto this external space, they become phenomenal representations—objects that the understanding must then discern and give meaning to. Hence, Kant's important statement: *"Thoughts without content are void; intuitions without conceptions, blind."*[45]

Time is also given to us *a priori*. According to Kant, we cannot imagine being outside of time, *nor can we imagine or give form to phenomenon outside of time*. Trying to imagine time coexistent (past, present, future), rather than successively, is impossible. We can imagine an infinite succession of time but not time simultaneously (for example, as posited by today's parallel world's theorists in respect to space). In as much as space is a most important externally expressed intuition, time is the most important internal intuition, subsuming even space. Through time we gain a sense of self and the internal state of our affairs.[46] Kant puts it this way:

> Time is the formal condition *a priori* of all phenomena whatsoever. Space, as the pure form of external intuition, is limited as a condition *a priori* to external phenomena alone. On the other hand, because all representations, whether they have or have not external things for their objects, still in themselves, as determinations of the mind, belong to our internal state; and because this internal state is subject to the formal condition of the internal intuitions, that is, to time—time is a condition *a priori* of all phenomena whatsoever—the *immediate* condition of all internal and thereby the *mediate* constitution of all external phenomena. If I can say *a priori*, "All outward phenomena are in space, and determined *a priori* according to the relations of space," I can also, from the principle of the internal sense, affirm universally, "All phenomena in general, that is, all objects of the senses, are in time, and stand necessarily in relations of time."[47]

In sum, Kant makes it clear to us that phenomena are subject to an internal sense of time but manifest themselves in an external sense of space. In other words, the *a priori* conditions that constitute our minds allow our sensible intuitions to consolidate all the pieces of the puzzle. They become the phenomena that we see in real-time as solid objects. This view is not unlike that of Dale Purves' thesis in respect to vision; he believes that we see what we see and recognize the objects we do because of centuries of accumulated pieces of visual information stored in our brains (see chapter 5). One can also cite neurophysiologist Karl Pribram's idea of a holographic brain, whose parts contain the pieces of the images that present us with a whole image.[48] In the *Critique of Pure Reason*, Kant also posits that the whole of reason reflects all of its parts. Pure reason can thus be likened to the dynamics of a holographic universe, as reason becomes a whole that is made up of many parts. By limiting appearances to reason, Kant rules out any possibility of sensibilities going beyond it. His notions of time and space, set against the parameters of reason, provide the essential background against which all phenomena appear and situate Swedenborg's otherworldly visions in the realm of fantasy.

Like Kant, Swedenborg also wrote extensively, contributing to two very diverse fields. He was, on the one hand, a very well-known citizen/politician who specialized in advising his nation on the practical aspects of governing a nation and wrote many articles on complex scientific subjects. At the same time, he was not shy about reporting his personal otherworldly forays; what some people today would consider to be "flaky" experiences—talking with angels, having out-of-body experiences, hearing voices, and so forth—hence founding a theology that involved but at the same time reworked the notion of philosophical concepts. In writing about man's exterior and interior states, Swedenborg included the ideas of heaven and earth. Of interest to us is how he conceived otherworldly spirits and angels in space and time. Since his views on these subjects differed from Kant's, it is best to consider, briefly,

Swedenborg's encounters with these appearances—spirits and angels from heavenly dimensions—which take on a tone somewhat different from his philosophical writings.

To begin with, according to Swedenborg, man lives a natural life on earth, during which time he has sensual sight that contains the seeds of spiritual sight. Only a few men, on rare occasions, see glimpses of the spiritual life while on earth—and Swedenborg claimed to be one of them. He professed to crossover into the spiritual world and converse with angelic and spiritual phenomena, which were representative of people who had died on earth or on other worlds, including other planets. Death was the vehicle that enabled a man to crossover and live in virtually the same environment in which he had lived before death, as it corresponded in every way to the heavenly world he found himself in after his death. The heavenly world was so like earth that it confused him when he first arrived. Most men refused to believe that they had died. Instead of the flesh-and-blood body he had on earth, man obtained "a human form that [was] similar to [his] affections [love]."[49] When a man died he remained essentially the same man, except that he shed his human body and sensual sight and received a heavenly, youthful, virile body for eternity; most important, he obtained spiritual sight. (An evil man was given the opportunity to enter into heaven and receive angelic instruction but usually chose to remain evil and go to hell.)

As a newly arrived spirit, man responded to the heavenly world according to the degree of knowledge and understanding he brought with him. This turned out to be far superior to that which he had known in life as a man. After instruction by angels, his new spirit was ushered into the heavenly hierarchal class to which he belonged—this according to his spiritual accomplishments while in the natural world. Swedenborg clearly intended his theology to be perceived differently from that of Christianity and its teaching about heaven and earth. For example, Swedenborg believed that angels were not made before men; rather, angels became angels because they had once been flesh-and-blood men, no matter where they originated in the universe.

Swedenborg saw his interior space and time as correspondences yet correspondences that were very different from the exterior space and time we experience on earth. Space and time, although the same everywhere in the universe (because of Swedenborg's law of correspondences), were not perceived by spirits in the same way as they are on earth. Space in heaven is not the same space we experience on earth, because on earth we are limited to natural vision. Space in the spiritual dimension is *appearance*, and unlike Kant's space, which is "phenomenal omnipresence because it connects all things,"[50] an appearance in Swedenborg's space is an interior experience, not an exterior projection. It has been interpreted as being dreamlike, as in a dream, space and time are usually irrelevant.[51] A spirit or angel experiences appearances according to his own level of spiritual perfection. This notion of "appearance" that displaces space can be somewhat difficult to grasp. By it, Swedenborg means to convey that space is perceived differently when in the presence of divine love and wisdom.[52] He tells us that even though heaven has a corresponding place and space to that of earth's, angels do not see their spiritual world through it, as we do while alive, but rather they experience it interiorly. Therefore, any measurements they make are made differently, through their achieved interior states.[53] Because everything in the mundane world corresponds to the heavenly one, the sense of time also falls into the same category as space. Because time is eternal, space can be interpreted as being ephemeral.

The heavenly hypothesis of space and time that Swedenborg offered can present problems to those in any epoch if they cannot transcend the ideas of geometrical space and time necessary to grasp or *feel* this notion of an interior sensual space and time. Swedenborg, as I interpret him, did not want to teach a transcendence of time and space by using philosophical concepts or mathematical models that prevailed in his day. Ostensibly, he wanted to purvey something sensuously heavenly within us, in our hearts, even while we were alive, rather than seeking it in the starry heavens as we see them above our mortal heads (although he used up/down themes for his interpretation of where heaven and hell were located). His visions of what being in a heavenly state was like were, for the most part, embellishments of earth (replete with eternal sexual virility), which is not surprising—few, to my knowledge, have been able to imagine images of heaven without natural environs or human beings in it. Angelic existence constitutes being in love and in truth throughout eternity and is interiorly felt. Swedenborg, therefore, classified his meetings with spirits and angels as interior events and not exterior visions.

Thus, Swedenborg's spirits and angels do not live in an Einsteinian spacetime; they live in a structured, hierarchical order of states (which I would refer to as potential or intellectual states of consciousness), with the last state being pure, blinding light. As spirits and angels, they reach a person by communicating directly with his/her mind, rather than through the natural world's language facilities. (Mind-to-mind communication has always been scoffed at by sceptics, yet today, it is interesting to note, scientists have developed a way to capture signals from the brain/eyes of paralyzed people that enables them to operate computers via their minds—a mind-to-computer communication. Scientists predict that this kind of communication with computers could become standard fare in the twenty-first century.[54])

Kant writes that the appearances recorded in Swendenborg's "'Memorabilia' are not 'visions' properly so called, but scenes beheld in the most perfect state of bodily wakefulness and which 'I have now experienced for several years'" (*Arcana*, 1885). Kant goes on to say that Swedenborg describes "two other kinds of vision which he rarely experienced, one as being 'taken out of the body' or reduced to a certain state between sleeping and waking: during his continuance in this state he cannot but know that he is wide awake."[55] These visions, Swedenborg claimed, differed from those of persons who were visionaries:

> Visions are often spoken of which indeed are really
> seen, but in phantasy. The spirits which induce such
> phantasies work upon persons of weak minds, and
> who are easily credulous; such persons are vision-
> aries, and the things which they see are illusions
> conjured up from outward objects, especially in
> obscure light. Visions caused by enthusiastic spir-
> its are similar to these but refer to subjects of belief.
> *Arcana*, 1967–68[56]

Apparently, Swedenborg did not wish to affirm the type of
experience Michael Talbot reported as happening to mystics
today who claim to enter into parallel and other dimensional
worlds.[57] Yet this seems to be what Swedenborg described as having experienced. Unlike Kant, who provided us with rigid laws
that describe a reason limited to earth, Swedenborg tried to open
up sensibilities to divine love, truth, and intelligence, claiming
also that the visions or "seeing" of spirits and/or angels were
even possible to open-minded people before death. Nonetheless,
his reported visions and his experiences in heaven confirm what
today easily falls into mystical categories of the external appearances, something he tried hard to disclaim. Whereas Swedenborg
allows for a subjective/objective dialogue, mind to mind, with the
phenomena or projections from other dimensions and believes
that they fill him with an otherworldly knowledge, Kant refuses
to acknowledge an otherworldly dimension where "things-in-
themselves" might exist and claims that all we can know is the
mundane world of phenomenon that we create from within ourselves—all of which are part and parcel of our minds and thus
are simply innate, imaginative creations of the mind. Recall these
already excerpted words of his: "On the other hand, because all
representations, whether they have or have not external things for
their objects, still in themselves, as determinations of the mind,
belong to our internal state...." Kant's reason held him rigidly

postured in a geometrical three-dimensional world that he could account for through reason alone.

This is not to say that the connections Swedenborg talks about—making mind-to-mind connections with otherworldly beings—could not have been caused by something like otherworldly cosmic electromagnetic interference with his brainwaves, which could trigger the holographic beings that he saw and with whom he communicated. As far as Swedenborg is concerned, he expounds a theology that has nothing to do with mysticism, but this does not prevent us from asking whether he is explicating anything more than what today we would call holographic images, which Neil Bohr has brought to our attention in our epoch through the dynamics of quantum physics.[58]

This question is no more than a prelude at this point and will be subject to a longer discussion, which we will address below and again later in the book. For the time being, we must keep Swendenborg framed in his own eighteenth century and recognize that his accounts of his spiritual adventures are heavily weighted by his dealing somewhat differently with Christian theological perspectives that have to do with the ideas of the resurrected body, space, time, love, truth, and divine ordinance. Swedenborg takes it one step farther than Christianity does. He believes that every religious perspective, no matter what its approach to goodness and truth, is represented in the worlds of the afterlife. Like Christian de Duve and others today, he believes that earth is not the only "earth" in the entire universe (something that Kant agreed with too[59]). Swendenborg states:

> That there are many earths, and men upon them, and spirits and angels from them, is very well known in the other life; for it is granted to every one there who from the love of truth and thence of use desires it, to speak with spirits of other earths, and to be confirmed thereby in regard to a plurality of worlds, and to be informed that the human race is not only from one earth, but from innumerable ones.[60]

Was Kant right, then, in criticizing Swedenborg's phenomena and the latter's explanation of his personal experience of other-worldliness? Oddly enough, we cannot ignore Kant's own sentiments on the subject of otherworldly phenomena. In respect to his ridiculing of Swedenborg, he writes the following to a friend (Moses Mendelssohn):

> It seemed to me the wisest course to take advantage of others and first do the ridiculing myself; and in this I have been perfectly frank since the attitude of my own mind is inconsistent and, so far as these stories [by Swedenborg] are concerned, I cannot help having a slight inclination for things of this kind, and indeed, as regards their reasonableness, I cannot help *cherishing* [emphasis added] an opinion that there is some validity in these experiences in spite of all the absurdities involved in the stories about them, and the crazy and unintelligible ideas which deprive them of their real value."[61]

Would Swedenborg's otherworldly adventures with spirits and angels be more palatable if we thought of the phenomena he saw as holographic in construct and thus part of a cosmos that can project holographic images at us or induce them within us? In Swedenborg's case (not unlike the founders of other religions), he had experiences of phenomenon that were as real as any real time image is. He had conversations with angels and received their instructions. Their messages, he claimed, entered into his heart directly and became part of him. According to Kant, all phenomena are explainable by the innate mechanisms of the mind itself, and Swedenborg was fantasizing. If Kant is right, then we must ask why men like Swedenborg put their illustrious careers on the line to write about conversations with spirits and angels from other worlds. But more important and central to our thesis is, what made Swedenborg believe in the reality of his visions? Was his visual brain especially gifted, enabling him to see them? And if we take him as seriously as believers do of any other religion's founder, were they projected onto space from within him or at him by the cosmos?

Today, neurologists and others studying the operative dynamics of consciousness choose to begin with the visual parts of the brain, examining how the physical nature of sight works. They investigate how the information we receive from the outside world and then process is projected as phenomena (although not too many study the neurology of how we process phenomena during dreams). Today's neuroscientists can take advantage of studying the brain by using MRI or CT scan digital equipment to create three-dimensional images of the brain's response to pictures or language. Unlike physicists, such as Niels Bohr, who claim the world of appearances is always in the process of being created by the observer who observes it, other scientists, such as bio-physicist Francis Crick and mathematician Roger Penrose, are more interested in consciousness and the brain, and they stay within these parameters. Hence, how the brain processes images, how and where neurons fire up in the brain, what consciousness is, and what unconsciousness is, all comes under the scrutiny of science. Unlike Kant or Swedenborg, however, scientists, for the most part, do not concern themselves with how or where perceptions or images originate in the first place, and that is why philosophers like Kant and Swedenborg and others are still important to consider when studying the visual brain in our digital age.

Worldviews and Spirituality

In our considerations of Swedenborg's and Kant's understanding of phenomena, we ought to consider that their worldviews hinge on their individual, innate capacities to understand and describe what phenomena is—and how they describe it. From Swedenborg's reports, we would have to conclude that his conception of innate projections or visions that go beyond this mundane three-dimensional world were part of a cosmic dimensional field of energy. Not unlike that posited by Teilhard de Chardin or Vallée, he was able to see these visions in an intensely subjective way, as though mediated through a cosmic electromagnetic connection. Kant, on the other hand, reported a starkly objective world. He could not accept Swedenborg's or any other mystic's affirmation of otherworldly phenomena. One can grant Kant the philosophical aptitude for being very adept at describing how we project spatial images onto the screen of finite spatiality, and, of course, for his insistence that all phenomena originate from our own minds. But we should remember that both men believed that phenomena and appearances in the natural world were the result of man's innate ability to project reflective images from an interior source onto the mirror of the outside world. Swedenborg, of course, added an interior divine source to the equation, which was supposedly triggered at the moment of death.

Cosmic Holograms

Thus far in the book we have often used the term *holographic* to refer to the types of appearances and visions people claim they have had, without our examining what is meant by holographic. It is time to examine the collage of themes proffered in Michael Talbot's *The Holographic Universe*,[62] mainly because some of these themes are the ones that will help us to consider a piece of the puzzle— to possibly throw light on the physical cause of appearances and visions. The dynamics of a hologram will be addressed in more detail in chapter 6. Suffice it to say at this point that when a photographer bounces a laser beam off the object to be photographed and allows a second beam to interfere with the light of the first, the result, when recorded on film, is a hologram.[63] While I do not believe with Talbot and others that this is necessarily a holographic universe, I do believe that the cosmos can project its own holographic images and pass along others that have been generated, perhaps from other planets or otherworldly dimensions that we cannot yet explain. Before we address why Talbot believes in a holographic universe, we need to examine from where he derives his convictions and for this, we begin by turning briefly to his analysis of the work of physicists Neils Bohr and David Bohm.

Bohm: Implicit/Explicit Orders

David Bohm's innovative work with the quantum model looks beyond the model of quantum physics that Neils Bohr, one of the founding fathers of quantum physics, proffered. At his radical best, Bohr believed it was meaningless to presume the properties of some particle before an observer had observed it.[64] His quantum theory was open to much criticism and debate, particularly by Alfred Einstein and his colleagues.[65] Bohr addressed the fascinating problem of nonlocality, but he did not pursue it.

Then, a student, David Bohm, pondered the problem of nonlocality, the seamless interconnectedness of things, where all points in space are equal to all other points. Bohm noted that electrons in plasmas behaved as interconnected wholes, even though they had individual identities. Although particles such as electrons appeared to be separate from one another and with individual wills, on a very deep cosmic level they were identical. Eventually, a breakthrough inspiration came. While viewing television one day Bohm saw an experiment where a drop of ink was introduced into a spinning cylinder of glycerine and thus was dispersed throughout it. When the spin was reversed the ink blob regained its original form. Bohm's insight was immediate. It occurred to him that when the ink dispersed into the glycerine, its original form was hidden, or implicit, and when it was reconstituted the form was revealed, or explicit.

For our purposes, we should emphasize the fact that Bohm's new quantum paradigm envisions a dispersed or implicit (enfolded) cosmic order that contains an explicit (unfolded) finite order, the latter revealing itself in the real-time. Bohm, by the way, does not believe that we create our reality by looking at it—by freeze-stopping a wave, so to speak—the way that Bohr does. An example Talbot gives here is of a holographic piece of film. When it is encoded by interference patterns, it contains only an implicitly enfolded picture of what the entire picture looks like when explicitly unfolded. In simpler language, when cut into pieces, each part of the film still contains the whole image. Bohm believed that how an observer interacts with the implicit/explicit ensemble determines which part of the hologram is seen. The whole of this implicit/explicit dynamic can be defined as being the holographic dynamic that rules the universe. This is the meaning that Talbot posits when writing about the holographic universe, and it is based on Bohm's theories of a holographic seamless fabric of the universe, where there are no separate parts and where everything is interconnected (referred to by physicists as nonlocality). We ought also to note that Bohm refers to this seamless holograph universe not as static but as a holomovement because he believes the universe is always in the processes of implicit/explicit motion.

Aside from Bohm, Talbot also cites and describes neurophysiologist Karl Pribram's theories. (Interestingly, both physicist Bohr and neurophysiologist Pribram arrived at similar conclusions separately.) Pribram's theory, which deals with the brain as holographic matter, allows for an organic liaison to the idea of a holographic universe. Talbot does not pursue the biological (genetic) aspect of the holographic universe; rather, he relies on Bohm's and Pribram's abstract theories as foundational for his own but overlays paranormal events on them. Bohm's and Pribram's holographic universe constitutes a physical theory that connects consciousness to the cosmos. With this model Talbot can explain paranormal and mystical experiences in a scientific way, in order to give credence to these experiences—experiences that neither Bohm nor Pribram are interested in doing.[66]

Nonetheless, Talbot's appropriation of the physics underlying the holographic universe allows him the formula by which he can access parallel universes or otherworldly lives; it allows him to easily explain the crossover boundaries from real space and time to other dimensional spaces and time—not unlike what Swedenborg professed he could do. According to Bohm, past, present, and future are nonlocal entities and are actually accessible to us but only if we relax, learn how to let go of our conscious constructs of sequential reality, and yield to a holographic, seamless imagination. If we free our minds of the limits imposed by our sequential time/space experience, we can enter into a holographic framework—into nonlocal, nonhistorical space and time. This is the same as saying that only when we free our minds of the illusion of the space and time in which we live can we soar beyond it.[67] Bohm's implicit/explicit dynamics, nonlocality, and the nonhistorical notions of space and time become strangely intertwined with Swedenborg's interior/exterior dynamics and what seems to be his attempt to describe a "seamless" space and time.

Holographic images are caused by interference patterns that are invisible because implicit, and visible because explicit. Again, Talbot provides us with an example: by throwing two pebbles into a pond simultaneously but within a close range and watching their ensuing ripples crossover each other, we witness an interference pattern. An extreme view holds that these images are always without substance and thus always illusionary. And while Pribram believes that our brains construct objects, Bohm's conclusions are even more radical: that we construct space and time. Talbot summarizes these theories: "Considered together, Bohm and Pribram's theories provide a profound new way of looking at the world: *Our brains mathematically construct objective reality by interpreting frequencies that are ultimately projections from another dimension, a deeper order of existence that is beyond both space and time: The brain is a hologram enfolded in a holographic universe* [italics in text]."[68]

The otherworldly reality and its inhabitants that Talbot adds to all this, which I refer to as cosmic images or appearances, are neatly connected by him to electromagnetic forces. The cosmos, as we know it today (and our present knowledge of our universe is limited to only a small part it), constitutes a vast energy source that is made up of electromagnetic cross-interfering fields.[69] If this is so, then cosmic energy is constantly interfering with our brain's electromagnetic fields. By crossing over into our brain's fields, it creates real-time images that manifest themselves to some people, who then, in less sophisticated language, give testimony to the eternal holomovement of the holographic universe. Phenomena that we could never explain previously are now explainable by this electromagnetic crossover physical theory.

Although Talbot does not make this quantum-inspired claim in an overtly philosophical way, later in the book I do, for I believe— from the gist of all the evidence that Talbot provides in his book— that holographic free-floating images, when observed by an observer, take on substance or matter when electromagnetic cross-interfering waves merge their particle forms to produce a holographic image. At first glance, this appears to be an extreme view, yet something like this may be true. Perhaps our brains actively participate in making the holomovement a reality in the spacetime that we experience. There is room here too to suggest that electromagnetic crossovers closely link Bohm's moving holographic universe to Bohr's quantum physics paradigm, hence becoming a holographic image, a "freeze-stop," when these dynamics are combined.

Parallel Worlds; Exotic Afterlives

In his discussion of the holographic universe, aside from the scientific perspectives and otherworldly experiences, Talbot recounts anecdotes that cite evidence for the existence of a paranormal terrain, which I will discuss in a later chapter. In the meantime I will stay with the evidence he provided on individuals who claim to have crossed over into parallel worlds.

Talbot writes that earthlings who have visited a parallel holographic dimension, or afterlife dimension—by virtue of the fact that they died but were then allowed to return to earthly life—say that they have been told by the "beings" there that the purpose of our life on earth is to attain knowledge of what it means to live an illusionary life, the dimension we call reality. According to Talbot, these witnesses, who have traveled to the otherworld, tell us that our prime purpose in life is to acquire knowledge and understanding while on earth, in order that we can learn about true love. Talbot describes dreams and out-of-body experiences that people have had of celestial cities that are breathtaking in their aesthetic construction of buildings, and that contain libraries with books of knowledge replete with information needed to attain ultimate knowledge and supposedly the truth underlying the holographic universe. The dreams cannot be properly described, he claims, because we have no way of depicting the profoundly strange beauty of these celestial cities through the construction materials we have available on earth. Talbot also describes how people like Swedenborg step over into a parallel but much more lucid dimension of what looks like our earth—it is similar to our earth, with vivaciously colored fields and flowers and magnificent scenery such as we have, yet it is in shapes that we do not have on earth. This account of a more luscious earth than the one we already have seems to confirm that the mind is limited to imagining images of earthly heavens and of biblical Gardens of Edens. Because the mind does this, it hints at the universality of transcendent archetypal but biologically produced images, whether they depict physical beings similar to ourselves or images solely scenic in content.

The kind of beings that one encounters in a parallel dimension, according to Talbot's appraisal of Swendenborg, do not chastise a person for her faults but rather helps her review her life. Apparently, one steps into a "movie script" of one's life and reviews and relives all the love, hate, and suffering one has experienced—and to which one has subjected others—but with much more intensity than one had in the original life experience. The purpose of such a life review is to help one chose a future life that will avoid all the pitfalls of the last life. Talbot's account of the holographic parallel universe is full of details of these and other kinds of psychic experiences, which he admits many people would just discard as superstitious nonsense. And finally, it is Talbot who cites the reference of Swedenborg, saying that "opening the Book of Lives was recorded in the nervous system of the person's spiritual body."[70] This was an insight that came to Swedenborg well before the genetic information we now possess and one that I pursue as relevant to proving that we are all part of the same biological cosmic material.

In presenting the reader with a thesis based on what is involved in life in an afterworld, Talbot suggests it is transcendence into another kind of knowing and another kind of reality that surpasses our supposedly mundanely lived illusionary reality. Therefore, whether we participate accidentally, or we knowingly enter into one of the parallel worlds' transcendent-like consciousness, we experience or partake in otherworldliness that is very different from our real-time reality.

One could surmise, by virtue of the underlying logic of the Bohmian holographic universe—or for that matter, Prigram's holographic brain—that everything in the universe would be the result of electromagnetic interference (what Talbot refers to as *"frequency domain,"*[71] and hence, that all parallel worlds—and not only the world we inhabit here on earth—would also have to be illusionary. Thus, what we perceive as our life on earth and our cosmos is only a holographic illusionary image. A scholar of philosophy and/or theology would be hard pressed to identify anything uniquely radical in this postmodern description of the holographic universe. What is radical is that we now know how to produce holograms. Most of these kinds of cosmic dynamics have already been described (but not as holographic) by the pre-Socratics, by Plato's moving cosmos, by mystics such as Jacob Boehme or Giovanni Bruno, and by many others. As Talbot points out, the ancients and even philosophers such as Leibnitz, with his concept of monad, have described such holographic illusion.[72]

More recently, in the twentieth century French deconstructionist philosophers such as Maurice Merleau-Ponty, Georges Bataille, Maurice Blanchot, Jacques Lacan, Jacques Derrida, and Julia Kristeva play with dual concepts such as visible/invisible, inside/outside, interior/exterior, the interweaving of flesh, invagination, and so forth.[73] When memes infiltrate global cultures, it is as though they are viruses pervading the air, everywhere at the same time. This phenomenon has existed since recorded history. Other examples to consider are the ancients' pyramids found across oceans on different continents but which were usually constructed at about the same time. Today, conceptual memes infiltrate academia very quickly, with scholars adopting enticing new dynamics to create or adapt models or formulas that will fit their own particular disciplines. Popular culture lags temporarily but follows shortly with its own material version of these concepts.

According to the foregoing chapters, then, and in line with some of Talbot's views, the projections of parallel realities into real-time experience (or vice versa) have been pretty constantly described over the years by our philosophers and theologians. Interestingly, over the past three thousand years or more, human imagination gives witness to the fact that it does not seem capable of changing the content of the psychic world of the holomovement it projects. In dreams or in out-of-body states, the kind of projections visionaries have—whether of UFOs or of the Virgin Mary—appear to be more or less fixed and consistent over the past two millennium or more. Visions of heaven and the blinding light beams that lead one to heaven; the accounts of appearances of heavenly bodies that are usually brilliantly white—these and all the other reported phenomena seem to be standard witness to the essential manifestations of holographic projections from the cosmos, whether emanating from within or without us. On the other hand, the physical analogues to these appearances that we produce—such as airplanes, spacecraft, three-dimensional art, virtual-reality productions, holograms, and so forth—confirm that our mental projections must be ineluctably bound to a lawful set of images that act upon us for a reason. Jung, as we noted above, described these types of images as emanating from outside of ourselves in his 1958 dream of UFOs. In the dream he saw perfectly circular lenses, with one metallic extension pointing or leading to a magic lantern. While still in dreamlike reverie, Jung thought about his dream's content and later commented, "We always think that UFOs are projections of ours. Now, it turns out that we are their projections. I am projected by the magic lantern as C. G. Jung. But who manipulates the apparatus?"[74]

In the next chapter we will consider how biological holographic images are projected. We will not leave behind the main themes found in this chapter; rather, we will carry forward, for continuity's sake, some of the ideas that Talbot and physicists have of a holographic cosmos in order to merge them with a biological concept of the cosmos. The question that will be paramount is what role our biology plays—or does not play—when the cosmos interacts with us by sending its projections at us, or we, by sending our projections at it.

I should mention at this point that the main difference between my thesis and Talbot's is that my considerations focus on and are based on a biological cosmos rather than a phenomenal one. The cosmos, as I perceive it, is one that is biologically substantive and interacts with our own biology. It is not really a new idea but rather a well-worn one, if one considers the early Greek metaphysics on the subjects of nature, soul, and self. "Germ theories" and "soul seeds" were posited by them, as well as theories of reincarnation, not necessarily as a human being but as any other natural form. Belief that there was a general pervasion of souls throughout creation that were equally distributed throughout nature, no matter what the substance, whether organic or inorganic, persisted throughout the ages and was repackaged for the modern mind by James Lovelock, who readdressed the ancients' beliefs of the cosmos as organism, with his Gaia hypothesis in *The Ages of Gaia*.[75]

CHAPTER 3

COSMIC GENES AND HOLOGRAPHIC PROJECTIONS

Dawkins: Blind Biological Projections

Having examined projection dynamics from Talbot's point of view, who writes about an illusionary, seamless reality that invites us to enter into a new understanding of the operative dynamics of a rather abstract, mathematical holographic universe, I turn now to a very different approach on a seemingly unrelated yet very much related subject—to Richard Dawkins and his book *The Blind Watchmaker*.[76] I do so because I want to shift our attention from the abstract ethereal to a biological one. Dawkins' biogenetic approach throws some very interesting light on how illusionary or holographic images might, indeed, be created out of our own biogenetic material. Indirectly, he gives credence to why archetypal forms are part of the *materiality of DNA* and why they are our projections and part of our *holographic imagination*. Hence, from a very different (this time biogenetic) perspective, Dawkins inadvertently, albeit in a small way, confirms the thesis of the holographic universe described in Talbot's book. Dawkins, of course, does not say that we create images that are part of the holographic universe but rather that we choose new forms out of the biological material of DNA—out of our own particular genetic endowment. Our genes, it can be deduced, are always striving toward the creation of more perfect forms.

Because Dawkins is intent on affirming Charles Darwin's evolutionary theory and clarifying some unrelenting misconceptions about the process of evolution, he comes up with an interesting model of evolution through the utilization of a computer software program. Dawkins designed *Evolution* to demonstrate the role that nonrandom selection plays in Darwinian evolution. Dawkins purports that the evolution of any species in nature occurs so slowly and over such long periods of time that people have trouble relating to Darwin and to what evolution means. Even "punctuated equilibrium" theories such as Steven Jay Gould's and Niles Eldridge's still are part of a gradual evolutionary process and thus belong to the gradualist theory category.[77] Ultimately, it is the body of the evolving creature that chooses the best possible form for itself as it adapts to its environment. What Dawkins seeks to clarify are the prejudices underlying some of the theories that decry Darwin's theory as being based completely on randomness. These points of view are false, Dawkins argues; evolution as described by Darwin is a process that is actually not blindly random but rather almost classifiable as intelligently selective—hence, the blind watchmaker.

Wish-Filling Genetic Forms

Dawkins devotes an entire chapter (chapter 8, "Explosions and Spirals") to explaining how positive and negative feedback work in respect to one aspect of Darwin's evolutionary theory, and he spends time elaborating on what philosophers or psychologists would categorize as an example of psychological projection. It is of interest to us because the gods and goddesses referred to in mythologies of the past generally have been believed to be projections of the mind in a derogatory sense. *Supposing, however, we thought of them as being the best possible archetypal choices of projected appearances for a developing human intelligence in the slow process of evolution?* Dawkins' example is useful, therefore, in demonstrating why a developing consciousness that can project archetypes should not be thought of as a hallucinating one. The example Dawkins uses is based on Darwin's successor, R. A. Fisher, and mathematical biologist Russell Lande's work on sexual rather than purely natural selection. We can summarize it briefly in the following manner: If female birds of a certain species are genetically programmed to prefer longer-tailed mates rather than shorter-tailed mates, then longer-tailed mates are destined to be chosen over shorter-tailed ones. Thus, a spiralling evolutionary cycle of positively reinforced genetic data is initiated and can proceed, if not checked, toward an exponential growth explosion. As Dawkins puts it: "So, when a female exercises her choice of male, whichever way her preference lies, the chances are that her own genes bias her choice and *are choosing copies of themselves* in the males. They are choosing copies of themselves using male tail length as a label...." [Italics in text][78]

The theory of a female bird being genetically programmed to choose an image of the ideal male plumage because it appeals to a certain dominate gene in her makeup as being ideal, Dawkins points out, can also be used as an analogy to demonstrate why human females might choose smarter men with bigger brains and thus reinforce the evolutionist theory of why the brain has swelled as much as it has during the last few million years.[79] Jungians, for example, would have little difficulty with this analogy because of their anima-animus archetypal projection theories, which they believe they have proven to exist psychically, mainly through their research into dream dynamics. Without a doubt they would say

certain women are attracted to males who are intellectually superior to other males, because their female genes lean toward a psychological need to complement their own animus instinct (gene).[80]

The point that Dawkins makes is important to examine because
it has to do with explaining biologically how genes, at a very
fundamental level, are involved in a self-developmental process.
Explaining the role that genes can play if females (of any species)
use their preferentially programmed instincts for choosing which
male to project their preferences on, females thus directly control
their species' evolutionary form. In Dawkins' example, evolution
occurs because the ideal form is somewhere out there in biomorph-
land,[81] waiting to be chosen by the "I." The form does not come
from some ethereal aspect of a Platonic cosmos; it comes from
within the person herself. If one isolates the role that sexual selection plays, then one could argue that the environment only provides the background noise in which the genetic projection needs
to evolve. Even in Dawkins' software program, it is his "I" that
designs and influences an evolution of insects by always choosing
the images that appeal to him. The computer simply provides the
environment of necessary data to formulaically develop the image.
So Dawkins' as well as Darwin's environments are limited to earth
or to an earth-like planet elsewhere in the cosmos. They do not
involve the cosmos itself.

But let us push Dawkins a little farther on this notion that a
female's genes are able to project her sexual preferences onto males
and hence control the evolution of the species. Despite his cau-
tionary note about using analogy, let us make the following one,[82]
based on what has already been said above. Is it possible that this
kind of genetic choice for choosing an ideal mate over some other
less ideal one constitutes a legitimate biologically working law, as
Darwin believed it did? If so, can we theorize that the appearances
and visions we have been discussing above happen in a similarly
preferential *projected* way? Do *seers* choose to see a particular
otherworldly phenomena over some other this-worldly phenomena

because of their own innate preference for certain ideal forms that they desire to be like? And is this biologically and cosmically programmed into their genes?

The Human Factor: the Eye and the "I"

The way that Dawkins' software program *Evolution* works is that it processes algorithmic input data and translates this data into an image on the computer screen. Given certain mathematical data, the program draws the appropriate matching image on the computer screen (for example, Edward Lorenz's butterfly image that appears when certain weather data is iterated[83] or Benoit Mandelbrot's strange attractors with their frilly borders, images that emerge when a mathematical formula is iterated). Dawkins reports that his own interaction with the computer becomes most important to the output. The human eye, as he puts it, is essential to his computer program, because the "I" has a tendency to select the emerging skeletal image that most appeals to it.[84]

The image that Dawkins chooses is indeed the one that appeals to him most; it is the one that appears to take the shape of an insect—or what his brain makes out to be somewhat like that shape. He reports to us that if he keeps selecting the shape on the computer most compatible to the insect that he *thinks* he is seeing, it begins to emerge. When he enters this selection into his computer program, he is rewarded with another series of mutations from which to choose and influence yet another shape he begins to favor. He writes that in his experiment, his tendency was to choose the shape most like the emerging insect he was imagining, although at no time did he have a strategy for evolving a specific shape—it was just, as he puts it, "a capricious" choice.[85] In this way he creates, after about twenty-nine generations of computer mutations, an actual image. He admits the images that result from his computerized *cumulative* evolution schemata floor him. They are, of course, the images of the insects that first emerged on the screen in skeletal form, which his eyes had unconsciously affirmed and thus reinforced. His "capriciously" selected choice of image had created the emerging images—and there they were now, these newly created insects, on his computer screen.[86]

As Dawkins concedes, the reason for his success in producing these insects on the screen, so like those we find in real life, is because *he* had a lot to do with it. He already subliminally knew which image he wanted his computer to produce. He had a mental picture of what was visually acceptable and what ought to be rejected in choosing one selection over another. Without his eye to affirm the choice and direction for the image, the evolutionary process he was creating on the computer would not have managed to produce the image it did. Dawkins even speculates that bees and butterflies might be able to create their own flower on the computer if the computer was placed outside and these insects were allowed to bombard the screen. After a certain amount of bombardments, the computer would print the result—an emerging flower of some kind.[87]

With the human eye playing the major role in Dawkins' *Evolution* software program by choosing and directing the computer image at every crucial point, is it possible to compare Dawkins' projection dynamic to the human or humanoid-type images that seers materialize out of an underlying instinctive imagination? Granted, Dawkins might not have known that this would happen when he began his experiment with his program *Evolution*, but now that he believes this does happen, that his imagination (or anyone else's, for that matter, who uses his program) takes on the role of creator, then it can be said that a person can create a unique image with appropriate computer software out of the genetic environment in his brain.

Although *Evolution* appears to deal with the potential to create all possible unknown images, we are not really dealing with all possible *unimaginable* images. The reason is, as Dawkins admits, that once he personally steps beyond the limit of what he knows to be a reasonable shape (it has to be like a shape in nature, if he is to recognize and select it), he reaches a natural dead-end. Pressed to the limit of his imagination, it would eventually fail him. Nevertheless, Dawkins leaves us with an insight into the understanding of how the brain, imagination, and computer work together to produce images in the way a sexually selecting organism might in a given environment that is also changing. This is important to note. Dawkins is no machine; he is a human being whose genes are doing the choosing and, in fact, mentally drawing out the form from the computer's software-encoded mathematical formulas.[88]

We now have another sense of how the finite dynamics involved in projection might work, one that is quite different from those described by the theories of philosophers and psychologists who have written volumes about phenomenal projections of the past. We saw how digital formulas implement Dawkins' choices in his program *Evolution*. We noted that the projected, unconscious-yet-consciously desired shape of an insect emerges from inside our brains and finds its way onto the computer's screen. What's going on here? What kind of interaction is occurring between human and machine? The analogy here would be that abductees are unconsciously choosing our future bodily shapes by the babies they claim to give birth to, and what Vallée refers to as a "spiritual system" is based on our own innate biological instincts—instincts most likely to produce the form we really desire to be. But our instincts could also produce a form we have already experienced in the past.

Let us consolidate some of the questions that have arisen and that we will attempt to answer in later chapters. If biogenetic projection is involved in the evolution of human consciousness, what role would images play? And who gets to choose the preferential images through which to build consciousness? Presumably, to use one of Dawkins' examples, our choice of imagery is the result of our natural evolutionary need to make our brains bigger and better. Yet if we are unconsciously choosing for a cosmically produced archetypal image, such as those discussed in earlier chapters, we still have to account for how these archetypal images that we project onto the cosmic environment got into our genes in the first place.

Addressing the evolution of consciousness is an almost impossible task because if Darwin is right, evolution proceeds very slowly. Yet Sumerian records, which suddenly appeared out of "nowhere" in history, seem to be saying that our consciousness developed exponentially. Seemingly, reams of knowledge just dropped from the sky for these particular people, supposedly brought to them by gods and goddesses. Could this have happened because of a positive feedback explosion—something Dawkins talks about? Can we ever know what caused the painfully slow evolution of humans on earth to explode overnight into an intelligent, literate consciousness? Yet another question arises in respect to the appearances and visions we are discussing. Do we project them because of a memory-laden genetic cache embedded in our genes and superimpose them onto the 3-D environment in which we are already immersed? If this is the case, are we limited to what we know when we project these images onto the cosmic screen, as Dawkins claims he was doing in his experiment? Is this why Vallée points out that appearances are limited to the imagination of a particular time? Is this why Swedenborg's afterlife worlds are so like earth and, concomitantly, why the heavenly worlds that a biblical prophet like Ezekiel describes are so earth-like? But we are somewhat ahead of ourselves at this point. We need to step back and examine another example of human projection, this one made without a computer.

Levin: On Strange Attractors

Having discussed how instinctively we can project the form that we prefer, we now turn to a further account of how a consciously or wishfully induced projection can be achieved. In an appendix to his book *The Opening of Vision*, a section he refers to as "Dzogchen Dark Retreat An Abbreviated Phenomenological Diary," David Michael Levin describes his experience in a completely darkened hut.[89] The experience has to do with the teachings of Dr. Professor Namkhai Norbu and the practices of Dark Retreat (*Yang Thig*), which constitute the core Dzogchen teachings. These teachings stem from Tibetan Buddhism and the Abhidharma and Madhyamika systems, which are Indian in origin. We will stay within the framework of what the author desires and not attempt to understand the doctrine that underlies these teachings; we will just accept that which the author relates to us about his experience in the dark for seven nights and seven days.

He tells us, step by step, what happens to him but particularly what happens to his eyesight in the dark. The first day and night were wondrous, as the play of lights that manifested to him were a delight to behold. He puts it this way:

> An incessantly changing display of forms kept me enthralled, entertained, and on the look-out: forms, like clouds, making their appearance, lingering a while, and then vanishing without any enduring trace. By the second night, I understood that this ceaseless play of light, this constantly changing display of shapes and patterns, sometimes suggesting familiar objects and fantastic landscapes, was a reflection of my state of mind. *The display was functioning like a mirror, showing me the inner nature of my mind.*[90]

What Levin tries to achieve through a series of "difficult visualization" exercises (which he does not choose to reveal to us) is to use the blackness much as one would a computer screen, in order to see what is going on in his mind. I must admit it is an interesting mental exercise, one that I had never heard of, but as I read on, I began to understand it in terms of what I was working on at that particular moment. It was very like the way in which the forms that strange attractors in chaos science generate themselves on the computer screen after a certain amount of iterations take place. When mathematical formulas or real-life data are fed into a computer and iterated, these formulas take on an image, a distinct shape or form; for example, the Mandelbrot set. Dawkins' computer program "Evolution" and the role of his own projection of insects places Levin's visual experiment in another light.

Levin reports to us a computer-like function of his mind that has a lot to do with the mechanics of visual projection, although he does not put it this way. He tells us that his eyes bounced around, oscillating wildly in the dark, to the point of extreme painfulness, because they did not have the normal visual boundaries or framework on which to alight. As he persists in practicing the exercises, which might include some kind of meditation (he does not reveal this), his mind is able to settle down and control, albeit unconsciously, the images that it is attempting to manifest. Eventually, without any conscious effort, Levin is startled by the manifestation of more than vivid colors. The phenomenon for which he has been fine-tuning his vision appears out of the dark as an "ornamental pelt worn by Sengë Dong-ma, one of the female dakkins and a supernatural being of light associated with the Dark Retreat teachings."[91]

Levin knows immediately that this is a vision manifested by his own mind, which affirms to him that there is no such thing as a "thing-in-itself"; that there is only mind and its manifestations. Levin's mind has produced on the screen of blackness an image of a "strange attractor," formerly bottled up in the recesses of his mind that was desperately seeking a way out. If we understand the points of dazzling oscillating lights that Levin describes as points in a blackness that are feeding back into the mind through the eyes, points that are always selected for form because of the "I" biases, it is not surprising that at some stage of his experience in the dark—and because of his effort to concentrate on the visualization exercises that he is doing—Levin is destined to create, out of his mind, the image of a strange attractor that is most important to his faith. Levin's visual experiment is quite unlike that of

Edward Lorenz's, who also saw a strange attractor in the shape of a butterfly emerge on his computer screen.[92] However, unlike either Dawkins' or Levin's projection experience, the butterfly was generated by a computer formula that was not affected by Lorenz's preconceived biases or projections. The emergence of the butterfly that took shape on his computer screen came from his accumulated weather data and was a totally spontaneous occurrence that came as a complete surprise.

Of the three projections, two are purposefully controlled by vision, while the third is strictly a by-product image of pure mathematical input. Levin's projected image is pseudo-archetypal because it belongs to the family of myth, rather than to that of scientifically programmed digital information or genetically programmed dream material. His mind has created the thing-in-itself that floats brilliantly before him in holographic magnificence. But since this thing-in-itself that appears in front of him does not exist outside of his mind and is also an important symbol of his faith, there is probably much more going on in this particular projection exercise than Levin cares to reveal. The minute that Levin materializes the holographic projection in his mind into space, he automatically affirms its archetypal nature (something we shall return to below).

Unscrambling Projection Theories

Levin aside, Talbot's and Dawkins' world of holographic and computer image projections may not be worlds apart after all, even though they seem to be. In Talbot's holographic world, because it is quantum-inspired, we can freely choose to project those archetypal images that are within ourselves and relate them to archetypal images that are outside of ourselves, and vice versa. In Dawkins' Darwinian world, we are presented with this very clear example of how we formulate our mental images. These are based on internally projected desires that create an externally evolving computer image, all of which depends on human/machine collaboration. In neither of their books do the authors lead us into related philosophical discussions concerning their theses that might invite us to deduce further insights from their work. Talbot remains within the abstract physical and psychical, while Dawkins stays within the biological and material. Their theories about projection become particularly insightful when we mutate them by cross-referencing their theses, so to speak, and addressing the results from both a biological and cosmic perspective.

Whereas Dawkins confronts us with a Darwinian world, where there is only a finite plane of consciousness, Talbot confronts us with many possible parallel worlds and thus different planes of conscious awareness in metaphysical dimensions. Because these models posit different dynamics when it comes to accessing space and time, we must be careful in differentiating imaginary dimensions of time and space from that of time and space that we consciously experience in reality. Biologists operate in a physical or biological framework of time and space. They require a beginning of flesh and blood that must take place in *real* space and time. Quantum-inspired physicists, however, have broken through these barriers of Einsteinian space and time by virtue of quantum laws, which in varying degrees, they believe, underlie the universe. I have cited these distinctions in order to differentiate between the organically based consciousness that Dawkins affirms and Talbot's worlds of parallel consciousnesses that function in an imaginary plane and operate metaphysically, because they are founded on electromagnetic, quantum-based dynamical schemata.

If Dawkins' analysis of his own experience with the computer-produced model of projection, based on his instincts in real-time, is any indication of how projection dynamics work, and if Levin did all his projecting of points in the darkness through his brain's visual abilities without the aid of a computer and screen, then is it not possible that any humanly produced holographic image operates, essentially, through similar biologically produced projection dynamics? We are already familiar with the quantum model that needs an observer to stop the holomovement and create an explicit image based on the holographic universe's scramble of images. We know by Dawkins' description of his own experiment that the computer's image could only be created with his "I." In both instances, images are predestined to appear because of the human being's unconscious predilection toward a specific genre of desired materialized projection, as Levin aptly demonstrated.

Talbot, too, underlines the freedom of choice of image in the selection and projection of the possible worlds lurking out there, ready to be activated by the mind. It is we who must select for the best possible image of an illusionary world. In a holographic universe, according to Talbot, all possible choices exist because of the free-floating flotilla of images beyond the space and time that we can tune into. It is we who project the possibilities for the future images of our universe onto this holographic screen because they are already programmed within us; hence, cosmic and material images shadowbox each other—the minute that one is activated, the other is, too. Here we have something going on that can be likened to Einstein's example of the identical angles of *polarization*—positive/negative charges that locate each other in nonlocality, although we do not know how they do it. In Levin's case, if the holographic image he projected was indeed as he describes it— brilliantly real—then might we assume that this materialization was activated by its polarizing complementary holographic archetypal form, even though it had to cross the boundaries of interior and exterior space?

Evolution, as Dawkins understands it, can sometimes proceed convergently. It seems to work on a pre-programmed blueprint that contains the best possible shape into which any individual organism will choose to evolve. Given the same conditions, another totally unrelated organism of similar biological construction that develops somewhere else in the world (or cosmos) will also evolve very similarly, because it follows an identical blueprint.

Dawkins gives examples of this phenomena of convergence by citing parallel biological convergences in organisms with unique characteristics, such as echolalia or electrolocation, which developed independently on the Old World continent and the New World continent and hit upon the same biological shapes necessary for their navigational methods to work.[93] Distant species that evolve in these convergent ways do so because biologically, when the organism is faced with the logical choice to optimize its shape and development, each proceeds in a convergent way—these are the right choices for it to make biologically. Dawkins does not take these convergent dynamics farther than the few examples that he cites. Today, we know that convergent dynamics exist on a much broader scale and extend beyond the animal species to include those of Homo sapiens—for example, the mystery of our seeming co-existence with Neanderthals.[94] Were these similar species part of a failed convergence? We will have to wait and see what kind of information the DNA extracted from the fossilized bones of Neanderthal man reveals to us, something that biochemical archaeologists are presently engaged in doing.[95]

De Duve: Cosmic Imperatives

Award-winning biochemist Christian de Duve, in his book *Vital Dust: Life as a Cosmic Imperative*,[96] confirms that similar convergent dynamics can be found in the biochemicals of cosmic dust, just as in those found on earth. De Duve takes his thesis of biochemical convergent choice deep into the heart of a prebiotic cosmos. He proffers that not only would these biochemical choices, made by the cosmos, optimize for intelligent life form, but they would do so exactly as they do on earth. The reason, he argues, is that biochemical dynamics pervade the entire cosmos. That is to say, if we were privy to a blueprint of the best possible pathway for chemical bonding to proceed in the cosmos (and de Duve comes as close to providing us with a map of this as it is possible to do, short of being the divine designer himself), he tells us that we would be astonished to see that there are only so many biochemical choices that can lead to workable biological constructs.

De Duve's thesis has since been vindicated. Some scientists believe that fossilized Martian bacteria was found on a meteorite with shapes that, apparently, "resembl[ed] some forms of fossilized filamentous bacteria."[97] When sliced thinly, chemical and microscopic tests of the rock detected organic compounds that could only have been produced by biological activity. The rock, billions of years old, was found in an Antarctic ice sheet. It is thought to have landed thirteen thousand years ago, and it revealed organic molecules called polycyclic aromatic hydrocarbons. The polycyclic aromatic hydrocarbons (PAHs) present evidence that primitive life existed on Mars about 3.6 billion years ago. To say the least, these developments have caused more than a stir in many academic communities. It is now fashionable to believe that the cosmos is seeded with life and that life may evolve anywhere at all in the cosmos, as long as there is enough water and light from a sun-like star to nurture it.[98] There is even evidence that one of Jupiter's moons, Europa, could contain the liquid water necessary to support life.[99]

De Duve also believed that there probably is convergent development in species throughout the cosmos, development that results in biochemical forms similar to our own being born. His thesis, that life is a cosmic imperative, overreaches Dawkins', which is limited to explaining how species evolve on earth. Yet there is a similarity in their approaches. In both cases, convergent developments on earth, or in the cosmos, proceed because of certain "best choice" imperatives for selection based on *nonrandomness*.

Can these nonrandom models be applied to the evolution of an *incorporeal* notion of consciousness? Is the evolution of such an intangible consciousness also based on a cosmic imperative? Is it possible that a humanoid species would always choose the best possible archetypal form, resulting in many such species developing in the cosmos, both bodily and mentally, like our own? Is this the reason for the worldwide cultural choices of the archetypal forms of god or goddess usually visualized as descending from heaven?

The further question then is, why these parallel archetypal projections? And why have they occurred in convergent ways all over the globe and possibly the cosmos? If Dawkins' evolutionary theory is correct—if the organism makes these designer choices for its body and mind throughout its evolution because of its genetically programmed ideals—then it would appear, because consciousness is also part of the biological process, that it, too, develops by selecting ideal forms, in more or less the same way that Dawkins describes it as happening biologically.

Campbell: Convergent Imaginations

We have, of course, many concrete examples of these kinds of convergent cultural developments regarding archetypal sacred projections. Outside of these archetypal projection examples we have other kinds of examples of converging parallels of worldviews through the material constructions of various cultures. The Mayas and the Egyptians show convergence in their hieroglyphic writing styles, which are not as diverse as they may appear to be—for example, both use animal-like figures plus syllables carved in stone, producing their unique iconographies that give witness to their worldviews and developing consciousness. They also show convergence in the structure of their pyramidal tombs and in the burial of their kings. Pyramidal structures also abound in the East, depicting the seven steps to the attainment of Buddhahood.[100] Between the East and the Ukrainian steppes we have many burial mounds that also follow, somewhat less grandiosely, pyramidal traditions. Along the same lines of converging and formalizing imaginations, we have convergence today in the style of the office towers we are building throughout the world, to say nothing about almost identical-looking airport facilities worldwide.

Convergent dynamics that manifest themselves globally demonstrate that diversely scattered peoples would not have developed similar cultural styles only because of self-organizing archetypal projections; rather, they would have developed in parallel ways because these innately pre-programmed genetic choices could not be avoided and because of the cosmic biological possibilities received electromagnetically by their minds, something we will discuss in more depth in a later chapter. Today's philosophers are busy arguing the pros and cons of Dawkins' "memology," which he proposed in *The Selfish Gene*.[101] Unlike genes, memes operate culturally as ideas and appeal to philosophers such as Daniel Dennett.[102] They catch on as a clothing style (for example, jeans) does in a population. Memes travel like abstract specters, infecting the public cross-culturally and cross-globally, and today are instantly transmitted via the Internet—the ultimate phenomenon of the late twentieth century that will probably reign supreme for some time to come.

Archaeological evidence proves that these kinds of stylistic trends happened in archaic times, too, when there was no Internet to set ideas electrically flying through the population, and people were separated by vast expanses of oceans and formidable terrains. Dawkins agrees that cultural trends such as these indicate that there is something "quasi-evolutionary about many aspects of human history."[103] But he is also aware that there is something more to the "meme" than just his original insight. He states about the meme:

> My own feeling is that its main value may lie not
> so much in helping us to understand human culture
> as in sharpening our perception of *genetic natural
> selection*.[104]

If brains are programmed to select for the best possible forms—and Dawkins argues that they are—then inevitably, similar archetypal images are fated to evolve naturally and cross-culturally, regardless of whether there has been any direct contact between them, culturally or historically. But this adds a quasi-evolutionary dimension to them. We see diversely unrelated cultures that have been destined to take convergent directions in respect to their archetypal notions, which then created their worldviews. As Joseph Campbell wrote at the completion of *Masks of God:*

> Looking back today over the twelve delightful years that I spent on this richly rewarding enterprise, I find that its main result for me has been its confirmation of a thought I have long and faithfully entertained: of the unity of the race of man, not only in its biology but also in its spiritual history, which has everywhere unfolded in the manner of a single symphony, with its themes announced, developed, amplified and turned about, distorted, reasserted, and, today, in a grand *fortissimo* of all sections sounding together, irresistibly advancing to some kind of mighty climax, out of which the next great movement will emerge.[105]

Campbell's long years of studying myth, resulting in his belief in the underlying unity of the race of man, has, I am sure, something to do with humans tuning in to the vibrations, to the resonance of the cosmos. We have even uncovered genetic evidence that links all races. It is out of this *cosmic biological dust* that images for ideal archetypes project themselves upon us, as well as probably on to any potentially intelligent life form lurking out there in the cosmos.[106] Within the biochemical nature of cosmic dust lurks the already pre-programmed best possible humanoid form out of which humans on earth were destined to evolve.

Plato, for example, came close to describing this biogenetic factor found in cosmic dust. He believed that souls were out there in the cosmos, waiting to be embodied in flesh and blood. Without a body, rationality could not emerge and a soul simply hovered somewhat aimlessly in the cosmos, waiting to seed itself on the appropriate planet. We have discussed above how de Duve believes there are many biogenetic forms in the vast regions of the cosmos that appear to be similar to our own. If this is so, then I suggest that cosmically produced biological archetypal images emerge from it. The Bible has it that God from somewhere out there in the heavens made humankind in His image. We cannot ignore this story—or other cultures' creation stories—for there is not only a sameness but a rudimentary sense of recollection going on in the memories of human beings with respect to the cosmos, influencing the evolution of earthly forms. Is the image proffered one that is a cosmic imperative remembered from long, long ago? Is it a message containing our cosmic origins but somewhat mythologically garbled because of time? The Bible reminds us in its first chapters that we are not capable of producing ourselves but are ineluctably bound to something cosmic, which we mostly refer to as God. While we may chose certain forms and then evolve into these forms through our projections, we also respond to received forms of cosmic projections, resonances of the mystery underlying the nature of our cosmic consciousness.[107]

Cosmic Archetypes

The same archetypal/cosmic forms that were the best possible forms and that made sense for the intelligently evolving consciousness in Homo sapiens to choose as *their* desirable form also manifested itself through the same genre of consciousness throughout earth. This is evident in every culture, ancient and modern, and all over the globe, particularly through our projections of the notions of divine beings. Joseph Campbell has drawn our attention to this phenomenon as it occurred in global mythologies, always more similar in story than diverse. These myths cannot help but uncover and confirm the dynamics of convergence, because of their projections of the same archetypal forms and because the same teleological goals can be achieved through them.

In studying Campbell's conclusions we can easily read into them Jung's claim that archetypes are instincts, although Jung did not pursue his ideas in a neuro-biochemical way, such as, say, Antonio Damasio has done in his book *Descartes' Error*.[108] Damasio, a neurobiologist, points out that instincts that run through our blood create the images necessary for the brain to respond in a normal, healthy way to reality. Only then, and out of these instinct/images, are concepts formed. We ought not to forget that almost all extinct cultures relied on human sacrifices in order to drink the blood of the victims whom they identified as being their gift to the gods. Blood transubstantiated into ambrosia, literally a food for spiritual growth. This flesh-and-blood sacrifice is recollected even today in certain Christian denominations that celebrate drinking the blood of Christ and eating his flesh during communion, albeit metaphorically.

If we add to Campbell's work on mythological consciousness the idea of the sacred's working from above down on humans, and Dawkins' Darwinian horizontal evolutionary theory, then we come up with something that combines them in the following manner: We have an archetypal/biological consciousness that, as it evolves, inspires diversely scattered people all over the globe to make convergent choices by projecting similar ideas/ideals of archetypal forms within a certain epochal framework. When we combine the mythological with the biological, we begin to see more clearly how convergences of cultural consciousness work historically. New scientific evidence about our biogenetic convergent ties to the cosmos not only brings all these different cultural and mythological perspectives together but invites us to ponder dimensions beyond the earthly, while remaining within its elemental confines; it invites us to wonder whether we are also somehow on the receiving end of resonant projections from the cosmos because of these convergent biochemical forms.

Do we make these evolutionary collective choices for consciousness because electromagnetic conditions or their resonances prevail throughout the cosmos, allowing some of us to tune in to the vibrations of a biogenetic universe that communicates to us through holographic images? By applying Dawkins' projection dynamics to the evolution of human consciousness, we also can begin to understand why, given the same cosmic existential conditions, we earthlings historically have come up with the same projection responses—some more emotionally based, others more intellectually, but all professing that something acts upon us from above.

We all can relate to the wonder of starry nights, the weather patterns, the availability or unavailability of food, the mystery of procreation, and the fact that all human beings are always guided by the irreversibility and unfathomableness of death. If this is biogenetically biased, then not surprisingly, the evolution of parallel religious practices were destined to spontaneously evolve all over the world, offering *parcels* of the truth about our own "divinity" that, as is argued here, was created by our own projections and boomeranged back at us from the cosmos.[109] From this we can deduce that human consciousness also appears to work in an evolutionary Dawkinsian/Darwinian way by choosing the best possible nonrandom progression in which for it to develop. It does not, however, do so alone—there is a background cosmic biological material acting as a cosmic imperative, guiding us to choose the best possible form from our innate genetic inheritance.

Converging Talbot and Dawkins

We have already noted that Talbot's parallel worlds are all part of a seamless holographic beyond the spacetime scenario—a holographic world that needs only to be activated by someone's projection of it to make it real. Dawkins provides us with a similar example that involves imagining parallel worlds in the process of Darwinian evolution. As he puts it, we must imagine "all real and conceivable animals as sitting in a gigantic hyperspace,"[110]as though having their own branch on a hypothetical tree. "We can think of most of this 'tree of all possible animals' as hidden in the darkness of nonexistence. Here and there, a few trajectories through the darkened tree are illuminated. These are the evolutionary pathways that actually happened, and, as numerous as these illuminated branches are, they are still an infinitesimal minority of the set of all branches. Natural selection is a process that is capable of picking its way through the tree of all conceivable animals, and finding just that minority of pathways that are viable."[111]

Dawkins, of course, would balk at the thought of past, present, and future worlds as intermingling in holographic seamlessness. He would balk at parallel worlds or afterworlds as retrievable somehow by human consciousness, because they are out there waiting to be brought into holographic focus by the mind. He believes in nothing but cumulative natural selection with this newly introduced notion of an "I"-controlled projection for the evolution of the best possible form. Like Hume, Dawkins would, I am sure, have nothing at all to do with miracles, out-of-body experiences, near-death experiences, angels, etc. Nonetheless, pushed to furnish us with some sort of conclusion, he, like other postmodern thinkers, provides us with a solution that can belong to any postmodern discipline, whether biological, mathematical, physical, or philosophical. The key words he uses are the same: nonexistence, hyperspace, projection, etc. This postmodern language can leave us with an image of a hyperspace filled with hypothetical trees on whose branches dangle all possible parallel worlds that could have materialized but did not.

All this brings us back to Talbot's holographic universe and his argument that every image we have imagined in the past or will ever imagine in the future will be forever imprinted on a holographic universal grid. So let us do the obvious: superimpose this holographic grid onto Dawkins' biological grid, which claims that all these biological images also are waiting to be actualized but that only the appropriate ones will be chosen to self-organize into the form they will eventually assume. In Dawkins' case, the choice would already be programmed into the genes and then projected and affirmed through physical evolution. In Talbot's case, the images or forms are out there, ready to be activated in the human being whose brain acts as a receiver, which is activated by electromagnetic interference and thus can tune into cosmic archetypal ideals. I think it is safe to say that both Talbot's and Dawkins' models are sustained—actually, overrun—by projection dynamics.

In Dawkins we come across the most vital part of the process of evolution in the person of the observer. The observer is the key to his argument that the process of evolution does not occur without a choice being made nonrandomly. The choice for evolving one way rather than another is not haphazard. Still, the shape is not formed because there is a form somewhere out there waiting to be filled; rather, it is filled because of this constant biogenetic-feedback projection of the "I"—feedback projection such as that demonstrated by Michael Levin's mind. Only one particular form is intuitively chosen, and that form is allowed to take another step forward, thus formulating its self-organizing, evolving image. In Talbot's book all possible forms (images) float freely in a holographic universe. They are, to be sure, not unlike Platonic forms, but if we modernize these Platonic forms, we can say that they are genetically embodied in cosmic dust and can be activated by the "I" when influenced by an electromagnetic projection.

Despite the fact that Dawkins only argues for nonrandom choice in biological evolution, his newly-arrived-at thesis that genes control evolution because of a pre-programmed preference that projects the best possible design (as already discussed, the plumage of male birds), he still cannot avoid the noisy domain of all other possible free-floating forms that never made it onto the branches of his self-organizing tree. It is as though Dawkins wants it both ways: he wants the image of the horizontal self-organizing organism, and he wants the quantum infinity image of the possibilities of the organisms that might have been, had they been chosen. I think this need for a twofold paradigm results from the fact that Dawkins has not solved the problem of where the innate forms, which any animal's genes prefer to project on the opposite sex, come from. Indirectly, however, Dawkins has already solved his own problem nicely. The solution lies in his personal experience of the overseeing of the emerging insect's form on the computer screen. The dynamic could not have unfolded the way it did without the subliminal design preference of the human being—his own "I"—along with the pre-programmed algorithms in the software that produced that image. The interaction between the two, human and computer, were necessary in order to produce Dawkins' computer insect.

Again, it is interesting to apply Dawkins' biogenetic dynamic to Talbot's collation of dynamics that make up a holographic universe—dynamics that hover around us, a mass of indiscernible noisy interference waves. Talbot's holographic universe's screen vibrates brightly, like a pixelated computer screen, waiting for a human being to activate it, as David Levin did the screen of his mind. The holographic universe's screen, like the computer's screen, also needs the gene preference selection of the "I" if the right image is to appear on the screen of real-time and/or on the screen of an alien's spacetime. Let us assume that Dawkins is right about the female bird's ability to control the male bird's plumage and that the human being's "I" is genetically programmed to prefer certain archetypal images over others. When we do this, we come up with images like those that Talbot discusses in his book. Again, the blueprint for image preferences is quite clear when we look at our creation myths, our fairy tales, and other well-documented phenomena, like UFOs, angels and/or other messengers from God, and the Virgin Mary's appearing to people (Mariology), to say nothing of the experiences of people like Joseph Smith, Edgar Cayce, and Emanuel Swendenborg, etc.

This is perhaps the right place to pause and consider the post-Crucifixion appearances of Jesus in the New Testament. One only has to do an Internet search of "Resurrection appearances of Jesus" to get a comprehensive list of where such appearances are reported in the Gospels. The story that takes precedence over all the others is, of course, the very first sighting of what appears to be Jesus by Mary Magdalene and another Mary by the empty tomb. He later appears to eleven of his disciples in Galilee, where he instructs them in what is referred to as the Great Commission. In various versions later written by his commissioned disciples, Jesus eats with them to prove that he is not a ghost, stays with them for forty days, and then ascends bodily into heaven to be with his Father. So how should we relate to this appearance of Jesus? The disciples have a difficult time recognizing him as Jesus; Mary Magdalene thought him to be a gardener. Jesus appears to them but not exactly as he did while alive.

The stories we are left with seem to describe these post Crucifixion appearances too vaguely and leave us with a sense that the disciples themselves found them to be quite mysterious, ethereal, and out of their range of understanding. These appearances are still largely debated by scholars in this field. Most important to our thesis is that Jesus, by his death, is transformed into an appearance seen by many of his disciples and becomes a riveting focal point for Christianity. Christianity just would not be Christianity without these Resurrection appearances of Jesus. It is important to note that only a select number of people were privy to Jesus' appearances. So, can we deduce that what these disciples and other witnesses to Jesus' appearances are telling us is really about their own abilities to project visions of Jesus in a resurrected form, albeit not always a form that looked as he did when he had bodily form? Or were the appearances of cosmic origin? However we look at it, these disciples had insights into a Christ-like consciousness that they had never experienced before and thus, they began an entire new way of thinking about the world that was revolutionary. They had had a breakthrough into what an ideal man should be.

Christ-consciousness aside, clearly Homo sapiens have chosen, time and time again, to project particular archetypal images rather than any others on their biologically created spatial screens, because this, ostensibly, is the right way—perhaps the only way— with which we can proceed the evolution of cosmic intelligence and consciousness. If evolution does, indeed, function teleologically to select for the best possible efficient design, then there is really no reason why the development of consciousness should proceed outside of the Dawkinsian/Darwinian teleological model we discussed above. If this is so, and given the fact that Dawkins is dealing with a biological model, consciousness must be inclusive in it. Biologically evolving consciousness must also be a nonrandom process, choosing its ideal forms from instinctual images. Ostensibly, we should take this farther and say that an evolving consciousness in the human being, because it is biochemically based, must also project its preferred archetypal forms onto the appropriate biological material, perhaps found on another planet in the cosmos.[112] It seems to me that Dawkins has inadvertently stepped over the limits of his own Darwinian thesis by this rather intriguing revelation of the role that genes play through their pre-programmed preference for choosing certain ideal forms over and above others. It has provided us with an interesting way to link the biochemical genetic material found in the cosmos' biochemical dust and with a way to account for the human "I" affirmation of it.

Cosmic Consciousness

Is natural selection for consciousness ineluctably bound to our genetic projections and expressed in our epochal selections of certain mythological and religious archetypes? According to Dawkins, it takes long periods of time for evolutionary traits to overtly manifest themselves; perhaps that is why we are still engaged with the same appearances and visions that *seers* have seen for thousands of years. As we have already pointed out above, archetypes are instincts programmed into our genes. Plato's philosophy, for example, circumnavigates what I suspect might easily reveal notions of a cosmic genetic inheritance, an ideal biological archetype that must choose earth for its form to evolve in. Many other philosophers, like David Hume, have flirted with the idea of the cosmos as a living organism and of course in our time, there is James Lovelock's much-cited Gaia theory. The early notions of Plato, Hume, and even the more recent work of Teilhard de Chardin lie outside the present epochal scientific range of biological discoveries and cannot be supported, as ours can, by current scientific paradigms.

Despite the fact that I am arguing that the universe is made up of cosmic biochemical elements, I do not believe that we should ignore the daring premise that underlies Talbot's work, nor should we ignore physicists' claims that the holographic universe and its electromagnetic projections of illusionary other worlds really does exist. Are all the images that are projected at us or by us, therefore, always materialized because of the interference of electromagnetic waves? Do we—or can we—feel these electromagnetic waves? Some advocates of the theory of parallel worlds advise that what we feel is tantamount to what happens to someone who loses a limb and then imagines feeling that limb.[113] Although we may not be able to answer questions such as these at this point in time, we should leave no clue to our cosmic inheritance untouched. The evidence we have inherited or that we find constitutes a viable piece of the puzzle.

The clue that best fits the holographic universe that Talbot describes comes from ancient India and aptly describes illusionary dynamics. Mircea Eliade put it this way: "As the myths of Indra and Narada show us, even to make oneself conscious of the ontological unreality of Time, and to realize the rhythms of cosmic Great Time, is enough to free oneself from illusion."[114] The Indian plan is to free oneself from the "I" and from reincarnations that fuel illusion and hence would put into play the holographic universe we have just been discussing. We live in time and as such, all time—even cosmic time is illusionary (Maya). The trick is to realize what illusion is, to rise above it, and eventually even above the cosmic Great Time in order to reach beyond the "I" into a time-less, blissful state: Nirvana. This reaching beyond the "I," as some Indian religions try to do, would, of course, negate even the holo-graphic universe. Abstract ideas such as these are especially attrac-tive to young adults, who in their quest to find Nirvana often use drugs to get there.

By juxtaposing Dawkins and Talbot, their views together sug-gest that something seamlessly holographic, which is also biologi-cal in nature, points to how our consciousness develops. We are physical beings who have substance, who feel, who think, and who live in real-time in a material world. The real world is, indeed, unfolding out of an implicit genetic order—it is projected by us in response to the images provided by the cosmos. For example, Einstein's ability to project imagery changed the course of the mathematical imagination and hence, how scientists envision the cosmos. But it is not as cut-and-dried as this. Whereas we used to think that the physical world was the primal world and our inner psychic world was subordinate to it, we now are beginning to realize that it is the other way around. Images that we thought of as being only part of an inner psychic world are now revealed to be part and parcel of the language of a cosmic world.[115] Jung, for example, coined the word "psychoid" to account for inner mental states that interact with outer reality, by which he meant something both psychological and physical—what could be referred to as uni-versal mind.[116] This was a very important step in the right direction, toward recognizing how mind and matter evolve.

In the real world of flesh and blood, the evidence is irrefuta-ble and points to people who are, indeed, subject to electromag-netic interference patterns that allow them to enter into parallel worlds and/or to experience what afterlife dimensions might be like. Experimenters (such as Timothy O'Leary) using psychedelic drugs have discovered and reported on the positive and/or negative effects of mind-altering drugs on the brain and have actually pro-vided us with a great deal of insight into what happens when states of consciousness are altered chemically. Yet I do not think our current evidence—that chemical interference alone holds anything but a small fraction of the answer—can discern the mystery of how cosmic interference works biochemically to activate the appearances and visions of seers. One marvels, too, at the insights into the workings of the mind by philosophers such as the Empiricists, Locke, Berkeley, Hume, and even Kant, for whom, let us not forget, appearance constitutes only the phenomena of the real-time world because we cannot know the "thing-in-itself."

Could it be that in a relaxed state of mind or during sleep we become receptive to the input of electromagnetic waves and hence, we activate or become receptive to a cosmically formal archetypal language embedded in our genes? We do not presently know the answer to this question but will explore it in a later chapter. Dreams constitute a symbolic language that is always on the leading edge of prodding us forward toward a greater psychic unfolding, providing cloaked clues to our unconscious or even conscious questions, often through absurd manifestations that we could never imagine in real-time. Solutions to our daily anxieties and other problems are sometimes answered in dreams, to which many prominent figures in history bear witness. And as we have discussed, in the unconscious or dreaming state, some people are more prone to out-of-body experiences. Evidence shows that people who have been abducted generally cannot recall these abduction experiences unless they are hypnotized into less conscious states. Edgar Cayce, for instance, had all of his clairvoyant insights in a state of self-hypnosis; his secretary and wife recorded them by taking notes. This alone ought to clue us in to the fact that the liberation of the imaginative mind occurs when the brain is freed of its shackles during less conscious conditions.

My claim is that the holographic appearances and visions that have occurred to people over the epochs while they were in either a dreaming or conscious state no doubt occurred because of the susceptibility of these people to electromagnetic interferences. But electromagnetic interferences can also simultaneously affect the brains of many people at a time—as many as seventy thousand in October 1917 at Fatima, Portugal, all of whom saw "a silver disc" superimposed on the sun. The disc spun on its axis three times and then suddenly fell to earth in a classic UFO "falling-leaf pattern," bringing the mass of people to their knees in either terror or awe.[117] We are, indeed, on the frontiers of exciting research when it comes to electromagnetic interference and its effects on the visual brain and conscious or unconscious states.[118] Inroads are being made, as reputable scientists from many related academic areas begin to focus their attention on discerning the innate mysteries still enfolded in gray matter.

Our journey through history thus far has constantly bumped into these ideas of a holographic cosmos clothed in the fashion of the day. Bit by bit, the claimed phenomenal revelations of our connections to the cosmos have taunted scientific consciousness. Over the years there has been much literature accumulated on related fantastic subjects that often are only the product of a storytelling imagination intent on manipulating the scientific knowledge of the day. Today, for instance, we have the New Age phenomenon, a kind of pseudo-religious movement that hopes to save the environment or, if not this, to become cosmic spirits and leave this world altogether for another better dimension. All kinds of unlikely and sometimes very naive people claim to be channelers, to travel to other planets in their sleep, or to be in touch with extraterrestrial beings. For the most part, we find in this New Age mythology bits and pieces of currently available scientific information that is cleverly (or not so cleverly) put together to create scenarios based on a modicum of truth. One gets the impression that New Age authors superficially knit together the scientific knowledge that they read about in order to create new mythologies, which are not a product of legitimate projections between the cosmos and their minds but instead simply relate imaginatively conjured-up stories.[119]

Despite the fact that the New Age movement throws a shadow on the scholarly study of the subject of holographic phenomena, it ought not stop academics from seriously investigating the scientific evidence that we have of extrasensory phenomena and other similarly related transcendent experiences that people have reported throughout the ages. We have to recover phenomenal information of this sort from all available historical sources, which means not excluding even what our present-day New Age mythmakers are intuitively imagining.

CHAPTER 4

CREATING NEW MYTHOLOGICAL SPACES

Thus far we have considered holographic projections from the point of view of people who, throughout the ages, experienced appearances and/or just heard voices speaking to them, loudly coming from somewhere outside of themselves. Of course, voices and images also can happen simultaneously; we should keep in mind that both seeing visions and hearing voices are closely related, as both originate in the temporal lobes of the brain.[120] In the previous chapter we examined scientific models and paranormal anecdotes. In this chapter I shift our attention: first, to vision itself in an attempt to understand how the visual brain projects images outside or inside of itself using biological material in a mathematical way; and second, to how vision can be subjected to trickery.

Neurobiologists on Vision

Orthodox neurobiologists who study the visual cortex approach it from the point of view of its working mechanics. For example, they examine how incoming information proceeds; through which pathways it proceeds (the ventral stream or the dorsal stream), and the different functions with which these pathways are associated. Neurobiologists are generally in agreement that there are five visual areas—V1, V2, V3, V4, V5—each specializing in one or two particular aspects, although neurobiologists are not always in complete accordance as to the exact function these visual areas and their dynamics play. Semir Zeki, a neurobiologist, assigns the following roles to these visual areas: V1 receives visual impressions that are formed on the retina. In turn, it is surrounded by V2, which processes black-and white images; V3, which processes dynamic forms; V4, which processes color and form; and V5, which processes motion.[121]

While Zeki remains largely conventional in his approach to the dynamics of visual circuitry, he is most interested in how our visual brains process and project the visual perceptions they receive, and he believes that by studying artists' works we see how their visual brains perceive and interpret the world, which he claims differ from ordinary people's brains. Beginning with a more traditional model of visual circuitry and adding his own theories to it, he explains how the visual brain works in respect to art. He proffers that projected in artists' paintings are the dynamics of their visual brain in action, an interesting theory in and of itself.[122] Among other things he has discovered through MRI scans of the brain is that if the image seen is abstract only, an elemental part of the brain is activated because the visual brain has no comparative data to compare it against. If what is seen is representational art (a scene with which the viewer is familiar), rational parts of the brain are activated.

Other neurobiologists, like Dale Purves, chairman of Duke University's Department of Neurobiology, are far more radical in their approach to perception. Purves and his colleagues begin with their own set of questions and approach vision not from its calculative, mechanical aspects but from an accumulated history of reflexive responses.[123] Purves, an "optical iconoclast," elaborates on a theory of vision that could be perceived as being established on a quantum-mechanics ideology, because it is based on spontaneously choosing one true image out of a myriad of possibilities. Our choices, founded on statistics, change the incoherent data received by our retinas into the recognizable world we know today. He believes that programmed within our brains and spinal cords are learned reflexes of responses to images that we have seen over millions of years, which allows us to figure out what the unintelligible bundle of impressions is, which we then decode and see as a particular object. When we see something, Purves claims, it is because our innate reflexes kick in automatically (he refers to this as a "knee-jerk reaction," over which we have no control).

Purves and his colleagues call their reflexive theory of vision an *empirical* theory of vision, one that has all the visual sensations encoded in the reflexes and is ready to be activated through visual circuitry. According to this theory, light is essentially ambiguous and as such cannot illuminate and identify an object. If that is the case, they ask, how can our perception of any object arise through the circuitry of the retina? Lighting, in fact, just helps to create the illusions and tricks of the eye that produce the illusionary objects we cannot account for. The only reason why we consistently identify objects correctly, we are told, is because we have slowly accumulated statistical information about what we have seen over the course of our evolution. Our visual system has an embedded reflexive data base; we see the world as we do because we have perfected the neural networks necessary to produce certain visual stimuli. We are the products of a Darwinian visual battle of survival of the fittest; we have been willing to adapt to visual change.

Where we have erred in our present theories of vision, Purves argues, is in our naive belief that perception is based on light falling on our retinas. But light falling on our retinas does not contain any information in and of itself; a photon carries with it no historical data. What we process visually is generally what we already know. We could never discern what is out there in the world in front of us if we did not have realms of empirical data upon which to draw. Purves points out that although this idea of empirical data has been around for some time, it is eclipsed by orthodox neurobiologists who confine themselves to mapping the brain's visual circuitry and hence stay within the boundaries of what we know about its anatomy and physiology. Despite fifty or more years of work on the subject of perception, Purves points out that neurobiologists working in this field still cannot explain what a visual perception is, using their model of *visual circuitry*. Instead these scientists use a mysterious representational logic to analyze the real-time world and acknowledge empirical evidence as playing only an incidental role.

To underline his claims, let us summarize Purves' scientific evidence for reflex theory and his argument that light can trick the eye and therefore present us with images that are simply illusion.[124] The reason this occurs is because historically collected reflex data projects the results for a particular object on the basis of what it has in its database, which responds only to light. This tricks the eye by creating an illusion. Hence, Purves' illusionary aspects are created by light's trickery. These illusions are stored by the brain and provide statistical evidence for recalling a particular image; the collective data creates the same illusion for people.[125] We will keep Purves' views on illusion in mind, as they offer us a newer perspective from which to examine appearances. Purves also points out that the function of the brain is not unrelated to the mainly reflex function of the spinal cord. He believes that thinking about the brain as *an engine of reflex associations* is a step in the right direction toward understanding not only how perception works but possibly how the brain works as well. By including the spinal cord, Purves automatically includes the archaic part of the brain, which I think is valuable to us in our investigation.

Ostensibly, Zeki and Purves do not approach vision from the same perspectives, even though neither neurobiologist is orthodox in his approach to vision. Both men are careful not to use the language of genetics, and neither of them considers the genetic basis of vision or what role genes might play in perceiving and remembering the world. In Purves' case, if we switched words on him and replaced his *reflex* theory with *instinct* theory, we would have a basis for saying he is dealing with the humanoid's genetic inheritance of archetypes (Jung's instincts as images). Purves admits to the importance of the biological in his theory, saying that future study on the subject must involve neurons and their connections in the brain *with* the statistical processes involved in vision.[126]

Orthodox theorists like Zeki explain visual mechanics from the point of view of the dynamics of the circuitry of the visual system itself. This leaves us with an impression that all input arises from outside stimulation, which is perceived and processed by the visual brain and then returned to the eye to be projected as the surroundings we see. It is we who have to make a giant leap and somehow imagine how we transmit this created image, as a movie projector would, onto space. How are we to understand this leap of image from physical matter into abstract space? Add to this the further complication that everyone else around us sees the same environment, and our role as *projectors of reality* becomes such a mystery that most of us choose not to think about it. Philosophically, the problem of perceiving the world often leads to solipsistic reasoning, a philosophical logic that has been around for a long time.

Making the Visible Invisible

From yet another scientific point of view, there are physicists like David Smith and his postdoctoral fellow Dave Schurig, both at Duke University, who evidently agree with Purves that the visual brain can be easily deceived. Through complex mathematics and the use of a metamaterials "fabric" which has a variety of shapes embedded into it, incoming microwave radiation and light can be bent in a specific direction. Smith and Schurig have managed to make a small object invisible by using metamaterials. In other words, they have tricked microwaves into creating an illusion that, visually, simply is not there at all; it is invisible.

But Smith and Schurig do not foresee anything more dramatic than this happening in the near future, although the pressure is on them to make this illusionary trick happen with much larger objects. They are, however, being funded by the Defense Advanced Research Projects Agency (DARPA), namely, the U.S. Air Force, the army, the navy, and others who hope to utilize metamaterials technology to hide their airplanes and ships. So far, these two physicists are not working with ordinary light waves but only with microwave electromagnetic radiation, which is a form of light and provides an easier entrance point than visible light to make an object disappear. There are others, however, like Graeme Milton of the University of Utah and Nicolae Nicorovici of the University of Sydney in Australia, who have used a superlens to bend visible light and make tiny dust specks disappear. Their premise is based on the fact that the intensity of the trapped light cancels the incoming light.[127]

No matter how small a beginning or how long it is before experiments like these can be practically applied to technological innovations, once innovators put the idea out there, it is just a question of time before some entrepreneur discovers a use for them and they become part of our world. Suddenly, H. G. Wells' invisible man, Marvel Comics' invisible woman (who bends light to make herself disappear), Harry Potter's invisibility cloak, and *Star Trek*'s concealed ship take on feasible possibility that aficionados of science fiction applaud. So a familiar question surfaces again—which comes first: the mythology or the science?

Having considered what neurobiologist Purves thinks about vision and its statistical processes that can create either veridicality or illusion, we can now proceed to consider appearances against this scientific perspective. Since appearances, as we are addressing them, fall into the category of illusion for most scientists, what can we deduce about appearances in relationship to Purves' theory that illusions are also stored as part of vision's statistical processes? If this is so, then the appearances (illusions) we have been considering thus far must have occurred over a large period of time to have become statistically wired into our bodies and brains during the course of our evolution. People must have been seeing these apparitions for a long time in order to have amassed such a body of statistically produced evidence.

Illusionary Tricks and Technology

At this point and before we continue to ponder how appearances might be produced, we need to approach perception from yet another tangent by examining other tricks of illusion and how they are created. More often than not, technological scientists and artists these days borrow from models of nature. By examining their methods, we discover that there is an interesting self-reinforcing loop going on here, particularly because the digitized cyberspace experience allows us to create imaginary or virtual worlds which, even though products of a digital linearity, underline unique ways to enter into nonlinear spaces, virtually and physically. In these virtual worlds we can learn, through hands-on experience, what a molecule or virus looks and feels like; how to fly a plane; how to win a war by fighting in a virtual war; and so on.[128] We can construct a virtual world, enter into it, touch things in it, and experience it in almost every way by entering into an interactive relationship with it, hence allowing ourselves to learn and understand a technological innovation without harming ourselves. This kind of hands-on education, through the use of 3-D real-time interactive scenarios, is already available to educators.[129]

Our techno-artists proffer an excellent indication of what is going on within the realms of the imagination.[130] We have only to look at the virtual realities that are produced by artists with the assistance of digitized computer programs to begin to understand why our worldviews, however much some of us may resist change, are destined to be radically challenged. So how are artists, through their technological creations, expressing their worldviews of what the future will look like? How are these worldviews going to affect reality as we have known it in the past and know it today? When we look at techno-art creations, should we become uneasy about what these artistic renderings of virtual realities are saying about how our imaginative constructs are destined to create our world in the future? For those of us brought up on religious worldviews, some older than two thousand years, the twenty-first century looms ahead as a real-time science fiction scenario, a scenario that continues to unfold before our eyes.

Before we examine technological projections that fool the visual brain, let us consider an example of a well-known phenomenon produced by the visual brain itself. Robert Teunisse, a psychiatrist at University Hospital in Nijmegen, the Netherlands, reports on the connection between vision and projected images. An elderly woman patient, whom he knew was quite sane, relayed to him that over a period of several years, she saw images parade in front of her—two-inch-high chimney sweeps carrying ladders. Knowing she was not crazy but actually "seeing things," the woman attempted to grab hold of one little man, who naturally vanished under her grasp. She knew it was an optical illusion but had no idea why it was happening; hence, her consultation with a psychiatrist.

A disorder of this sort, where normal people hallucinate, was known and first described by a Swiss philosopher Charles Bonnet in 1760, after whom the condition (Bonnet's syndrome) was named. Reading up on it, Teunisse noted that nearly all the reported cases dealt with older people with impaired visual states. In researching the condition further, he examined 505 visually impaired patients and found that at least sixty met the conditions of Bonnet's syndrome. All had visual handicaps that ranged from glaucoma to the breakdown of the macula, a small area near the center of the retina.

All the people knew they were hallucinating, and all were mentally sound.

What, then, caused the hallucinations? Teunisse presents us with two different scientific perspectives. The first is that the brain produces spontaneous images that are not suppressible by vision, so we have something that proceeds outwardly from the brain to the eyes. The second is that damaged cells in the eye create non-sense imagery that is then interpreted by the brain. This second scenario would mean that the imagery originates from the retinal stimulation of visual nonsense and proceeds inwardly to a rational conclusion of what that nonsense might be. At the moment, no scientist is willing to speculate on what it is that might be happening to cause these illusions.[131]

This is an example of an interesting sort when it comes to vision and the brain and projections, because the example can be related to the historical attestations of apparitions seen by very well-known people, such as Goethe.[132] The first possibility, taking Teunisse's examples a step further, speaks to the activation of stored images in a statistical database. The woman who sees two-inch chimney sweeps walking around with ladders might be recalling, through these projections, storybook images stored by her brain when she was a child. The second possibility suggests that the woman's brain might choose to process the visually garbled data by zeroing in on that part of the visual brain and its stored memories that contain memories most closely matching the garbled images it is receiving. Either scenario works if the visual images have originated from a stored database genetically arrived at or added to through experience. I am inclined to approach this projection disorder from the point of view of output or projection by the visual brain, with light setting off the entire event as it reaches a diseased eye.

I think most of us can attest to the fact that a thought or image of someone or something can be activated by a quite unrelated incident or thought—one sees or thinks of something in particular and suddenly, an entirely different, unrelated thought pops up in the mind. A popular example that is often given is the triggering of a song by some unknown phenomena. The song can then continue to play and replay in one's mind for quite a while until finally fading into the recesses of the mind forever (one hopes).[133] In the case of a song, one does well to pay attention to its words because generally, they are calling our attention to something from the deep unconscious that needs to be addressed.[134] Dream images speak to the same kind of activation; they usually are related to a deep unconscious need and are presented in a symbolic language that needs to be deciphered before it can be understood.

The images that visually impaired old people experience, which are not hallucinatory but seemingly stem from the brain and the visual system, constitute an important scientific clue as to why phenomenon such as UFOs and apparitions of the Virgin Mary continue to be seen. In the visually impaired elderly, it can be answered by biology and stored image data. Meanwhile, as illusionary projections such as these continue to be researched from a biological perspective, we have the advent of virtual reality inventors who are engaged in creating phenomena to provoke our vision and fool our brains. In the same issue of *Discover* magazine in which Teunisse reports on vision and hallucination, we also have an entire section on the breakthrough of scientists working with light to produce virtual reality phenomena. One such report is of a prize-winning concept that produces "phantom projections." The innovator of the projected three-dimensional objects in space, Susan Kasen Summer, states: "The images appear so real that they defy the brain's ability to perceive them as virtual."[135]

So how is her ingeniously suspended 3-D object in space produced? By using a system of mirrors, beam splitters, and lenses in more or less the same process needed to produce a hologram. The special trick Summer uses is to project the object from behind a booth, rather than allowing the light to first reach the eye as it bounces off a flat surface. Summer can take a flat, two-dimensional motion picture of a person walking along in any setting and project it into free space as a moving image. Then, through the use of techniques, such as bending the image slightly and projecting it on a separate background, her images take on a solid-looking 3-D form. Her invention is now used commercially.

I have cited this example because Summer's invention uses lasers or manipulates light to affect the visual system; there are many other examples that could be used but this one offers a modicum of explanation from the point of view of science as to what might be happening when people observe UFOs and what might be happening when people report having encounters with aliens or being addressed by a God who hides behind a burning bush. Here, the two possibilities—output from the brain and input from the environment, especially in respect to 3-D real-time images superimposed onto real-time environment—can be compared to those images reported by seers or abductees. Purves' research offers evidence that the visual brain produces illusions that are processed and stored by the brain. We have also examined the innovativeness of our scientific inventors and seen how they can produce 3-D images or other materials that can manipulate and seduce vision into believing it is immersed in a three-dimensional virtual reality that aspires to be actual reality. These inventors appear to be duplicating the tricks or illusions that the visual brain is capable of seeing spontaneously.

Holographic Reality

Since virtual reality's tricks on vision, as crude as they may be, can be achieved by technologists and scientists, it seems plausible to proffer that our visual systems can function on two levels. On the one hand, we can automatically process the visual aspects of the environment we happen to be in; on the other, we can project holographic images upon it just like innovator Summers does with her holographic phenomena. If technologists and scientists can beam images onto our environment that appear to be real, with apparent substance, so that we have difficulty discerning between a flesh-and-blood person and a projected one, why is it not possible for the visual brain to do so? An appearance of this sort is no longer in the realm of science fiction but must be seriously pondered as a possibility. Witnesses to these otherworldly encounters throughout history have been constant in their descriptions of the phenomena they have seen, even though they have always described them against the environmental scenarios of their own epochs.

Technology proceeds so rapidly today that it is difficult to keep up with. Surely our digital times are an example of what Dawkins refers to as an "explosive event" in the history of evolution. As representative as these examples are of what kind of holographic images are being invented, it is really a mise-en-scène for the two kinds of biological dynamics we are considering: images that emanate from our visual brain and that we project upon the screen of our environment; or images that some cosmic illusionary projection projects at us. However it is that these appearances occur, technologically we are confirming, from very diverse scientific perspectives, that our so-called mythological projections of gods, goddesses, God, UFOs, the Virgin Mary, and so forth might not be just a figment of our imaginations after all; that there is a greater, very real dynamic underlying all these processes that reveals itself through the technological innovations addressed above—dynamics that are peculiarly similar, whether they are produced biologically or technologically.

Because of our current ability to technologically create virtual realities, we are now in a position to understand more fully those early accounts of what our forebears witnessed in ancient times. How long will it take us to break out of the notion that all these ancient projections were not strictly imaginative mythological projections but have to do with more complex cosmological/biological dynamics? Probably as long as it takes us to breakthrough and discern the God spells that still rule the masses on earth. And so again we ask, why have these God spells so dominated our lives in the past, and why do they continue to do so, if not for a valid underlying reason? Is it possible that they dominate our worldviews because they are genetically embedded in Homo sapiens through cosmic microbial particles that once found their way to earth's receptive and fertile environment and from which we have evolved?

On the Wings of Myth

Both technologists and mythologists generally categorize myths as being part of nonlogical thinking. Mythological thinking is supposed to have evolved in primitive minds that could not reason logically about their world. Nevertheless, in a surge of unprecedented primitive imaginative creativity it ingeniously concocted amazing stories full of technical-sounding apparatus to account for their origins and their being in the world. Analogies to today's test-tube babies were perceived and recorded, both by Sumerians and Mayas, as having been created by gods and goddesses; gods were perceived as having arrived from somewhere in the heavens (in heavenly boats, etc.). If we follow Julian Jaynes' thesis here, it seems that the primitive mind was able to imagine and to be privy to speaking images coming from the right to the left hemisphere and subsequently to imagineer a fantastic worldview on which the foundations of the concept of God were built. These worldviews continue to thrive in the many nations that are still ruled by this concept of God.[136]

This, then, is what I mean by the nonlinear, nonscientific thinking in the past. This is how mythology is still categorized by science, which nonetheless *faithfully* believes in the big-bang theory. Paradoxically, however, through the auspices of science, these once believed to be superstitious or hallucinatory mythological scenarios are becoming technological realities in our world. We, too, create test-tube babies; we, too, have gone to the moon—to say nothing about our exciting fly-overs of distant planets, particularly by the space probe Cassini, which sent a smaller probe, Huygens, down to Titan, one of Jupiter's moons,[137] and the launching and landing of the rovers Spirit and Opportunity on Mars, which, for longer than we thought possible, climbed up and down its hills to give us a panoramic view of the environment—a Martian view that makes us feel, eerily, quite at home.

But why are these technologies, once believed to be a product of archaic minds, still with us? Why are they giving life to new myths? And while all this is happening, why are people still reporting projections and visions, unless these visions are part of our brain's genetic inheritance, meant to remind us of how cosmic evolution proceeds and what future goals we should pursue. It used to be said that archetypes, if they exist, lived in the primitive or limbic part of the brain stem. But how this silent language of archetypes got into our brain stem is still a mystery, even to experts who believe they have broken its code, such as the Jungians do.

Science believes this symbolic archetypal language is hogwash, that as soon as rational consciousness evolved, as soon as we began thinking out of our neocortex's left hemispheres, we had the tools to step out of this superstitious existence of "virtual reality." Of course, this has not happened. We continue to live in an age of reason that has created the technological worldview we live in today, which is, nonetheless, still undergirded by archetypal, mythological symbols—strange for a twenty-first-century reality that is definitely not supposed to support the illusionary but rather to be very scientifically biased. In this twenty-first-century scientific reality, all hypotheses, if true, must always be testable in the real world, which means that all apparitions and visions need to be verified, if they are to break out of their mythological categories. And right now, like the old lady who could not grab hold of the very "real" chimney sweepers, we cannot grab hold of our very real apparitions.

According to reason, the scientifically approved world is the only world we should believe in, because it can be verified—all except for the big bang, which, as noted previously, must be taken on faith.[138] Although science controls the affirmation of what is real and what is not, especially now, because of the advent of the power of digitality, scientists are, ironically, producing inventions that fall into the category of déjà vu. These inventions reincarnate the mythologies of the past and reinforce the world of the archaic imagination. This is the world that, as we noted above, is still locked in past memories and is not necessarily located in the baser parts of our brain stems but in the DNA running throughout our brains and bodies. True, the primitive brain is that part of the brain that controls our automatic instinctive functioning, while the neo-cortex controls our thinking processes, but this is too simplistic a statement about our brains, which are very complex. Today, our knowledge about the brain is much more intricate than just a topo-logical mapping of it. Scientists are achieving remarkable break-throughs and are able to map genes in the brain and discover which ones are turned on or off.[139] They also are finding ancient viruses in our genome; namely, retroviruses that they think infected our ancestors millions of years ago. And if these have survived in our genome, it is probably safe to assume that much more genetic infor-mation, especially from the new research into the epigenetic phe-nomena, will be mined from the genome in the future.

If we stay with the larger view—that a biological transcendence from a primitive brain occurred to produce a modern brain at some point in time past—then it is interesting to speculate on whether the digital dynamic in which we are presently caught up will affect the brain's anatomy again. Will the brain evolve yet another evo-lutionary part, possibly even in a discernable physical way, that is more than conscious—that is meta-conscious—or has our brain found its finished anatomical form? And if it has found it, does it matter to our evolution? Is the brain's continual rewiring all that really makes us evolve in intelligence? Archaeologists have done a lot of digging through layers of soil to provide us with examples of artifactual handiwork and other artistic endeavours that enable us to appreciate how the primitive brain worked. What we see most in these artifacts are geometric patterns, which appear to be the same throughout the ages, that still constitute the world and upon which we base all our mathematical/geometrical calculations, even today.

In the twentieth century, newly found cave drawings in France provoked our curiosity because of their fluidity of form, rather than geometric content. Similar drawings have been found in caves in Spain, and all these cave drawings are thought to be anywhere from twenty thousand to thirty thousand years old. When these drawings are contrasted with the geometric patterns that are generally found on archaic artifactual culinary objects or used as designs to decorate dwellings, do they necessarily represent another kind of experience of the world? Since both examples of art prevail, I think it is safe to say that for a long time there has been nothing new about the brain's spatial or reasoning regions, based on what our archaic artifactual evidence confirms. These regions have apparently co-resided in the brain for a long time, although they may not have developed simultaneously or at the same rate.

Since evolution takes place so slowly, we cannot know whether our brain has stopped evolving in its physical, intellectual, or intuitive development. But this really does not concern us, as we are interested in the genes already present in the genome and their cosmic origins. Although genetic science has a long way to go in deciphering the role that genes play in appearances, I suggest that our genes contain the links to our ancestors' past experiences in respect to appearances. It is our genome's genes that will confirm whether the gods and their space vehicles were the appearances that, in the archaic worldview, represented a virtual reality to the people living at that time. Is it reasonable, then, to assume that the new technological experiences of virtual reality that we are presently undergoing are about to launch us, if not into an archaic spatial understanding of our world then perhaps into a breakthrough in the understanding of appearances? Do appearances work out of the same spatial dimensions in our brains as they did our ancestors'? If in ancient times appearances intuitively nudged reason toward cosmic exploration do they also launch today's scientific prowess in similar directions?

Since our present worldviews were constructed on the wings of myth, can it be—because of our new technological talents that embrace nonlogical (or nonlinear) thinking—that we are on the brink of researching how we could remember all these past experiences in a conscious way? Are we in the process of uncovering the meaning of the mythological underpinnings that have always been foundational for our world? Since our mythological worldviews in the past shaped a cosmic worldview for the entire peoples of the world, will the technological reality that underlies our Western world change the way we perceive our worldviews today by breaking through the religious and scientific mind-sets that rule our imaginations? But more than that, will myth reintroduce us to a cosmic worldview of an entirely new order?

Is a breakthrough like this really happening? Are we in the West about to enter into a brave new cyber age, only to see the old myths reappear in new clothing in what could be called an eternal mythological recurrence? Will there continue to be interference from the gods? Or will we become the gods and, like recurring dreams that stop occurring after we have deciphered them, will the appearances stop appearing? As we read our ancient culture's stories—their tales of hearing, seeing, and talking with the gods—are we ready to concede that these cultures really had no choice but to bow to some kind of *survival instinct that has to do with cosmic biology.* We might wonder what our cosmic genes are programmed to remember and reveal. Are they programmed to never forget remnants of their cosmic past; programmed to continue to respond to it? Is this perhaps the key role in the evolution of human consciousness; a hidden knowledge transmitted by holographic images in order to keep us evolutionarily active? If this is the case, are we ineluctably bound to a destiny that has us pre-determined to continually recreate new versions of the same myths in order to reinforce the same cosmic scenario, if so, why the redundancy, for what reason?

The Dynamics of Seductive Space

At this point I want to examine the dynamics involved in the making of new mythology in our epoch. To do this we have to consider the various approaches to images today. In the Western world in the past century, we have been enamoured by the seductive nature of what is often referred to as iconic imagery. Films, then television, and now the Internet have taken over the way we fantasize about life. Images once again begin to vie with words as the preferred mode of communication. Whereas in ancient times one could see moving and speaking images only through the performances of actors on stage, today we create virtual realties that continue to extend our ideas of how we want to communicate and to be entertained. Entrepreneurs are hotly pursuing the new comer on the block in the interactive market—virtual reality modeling language.[140] It appears that people do not want to passively watch films or play video games or even chat in Internet chat rooms any longer. What they really want to do is to enter into the scene itself by becoming part of the action as actors and create the storyline as they interactively affect the plot by making one choice over another. While the implementation of this sort of interactive virtual reality might have loomed as science fiction twenty years ago, the imagineering of this interactive world has already begun, as popular Internet sites such as Facebook and Twitter offer a means to live a form of "virtual reality." But is this just a prelude to a real virtual reality that we are about to create?

In ordinary reality, we interact with real people in real space and through Internet space by the use of cameras. We see people's facial expressions, watch their body language, hear their words—all this while they experience us in the same way. Someday in the near future, we will simply be represented by our holographic selves and projected into a 3-D scenario through the medium of cyberspace, where we can interact, speak, and walk around as though we were embodied people in real-time.[141]

This future virtual reality holographic experience of another person—as someone we can see and feel but who is not actually present in flesh and blood—will project us into an interactive experience that can, for the moment, only be described as dream-like. True, in a holographic real-time experience one would know that somewhere, in an actual place, the person with whom you were visiting in virtual reality actually existed; that his surroundings existed, even though all either of you were experiencing was each other's simulacrums in one person's space or in some in-between space. Is this future holographic possibility of face-to-face meeting in cyberspace that much different from the face-to-face meetings that a shaman has with his ancestral spirits, or those that our biblical prophets had with God, or those reported experiences with gods or aliens from other worlds, or those of UFOs, or those of fairies that our collective mythologies recount? Were these appearances simply projected from archaic minds onto the screen of spacetime? Was all this just a product of these peoples' projection abilities? It always astonishes me that human beings in ancient times could have been so creative—that they could have had such an imaginative capability that was so staggeringly inventive.

This said, let us suppose that we are taking part in this holographic real-time virtual reality and are projected by an electrical transmission through cyberspace. While we have not gone anywhere, our holographic body has traveled, revealing our bodily image to someone oceans away, as well as revealing her image to us. We walk with her, talk with her, and feel her viscerally, but this "her" is only a virtual simulation. Think again about so many of the experiences adolescents and adults have had in the past with gods or angels, to say nothing of what visual anomalies may underlie experiences that occur today to schizophrenics and those with bi-polar disorder. Here, we must ask what these visionaries actually experienced: reality or virtual reality? Joseph Smith is the most famous recent historical case. His encounter of two personages, one of whom was the angel Moroni, occurred for the first time under a tree on a hilltop near his home in upper New York State [142] The same angel Moroni appeared to him a second time and on many occasions during the night, each time, as reported by Smith, in exactly the same way and conveying the same message. Think tape recorder; think movie projector; best of all, think a virtual reality projection of some sort, holographic in nature. If this was so, how was it projected into Smith's bedroom in real-time? Could Smith have seen a virtual reality that was taking place somewhere else in the cosmos in someone else's spacetime? Or was the angel Moroni simply projected by Joseph Smith's mind as a superimposed image onto his environment in much the same way as Levin produced the glimmering belt out of the darkness?

These considerations become even more interesting and provocative when we consider them from an alternate perspective, as cosmically generated. We are entering an era where bodily excursions into a virtual reality (as in a simulation of reality, such as in a flight simulator or as in an immersion into 3-D art) are going to become part of our everyday experience; holographic projections of teachers are already used in some areas of education.[143] We are poised to communicate through a medium that allows us to be somewhere else in a virtually real body and virtually real spirit. Like ghosts in a machine, we will float around in cyberspace and communicate with each other face to face, body to body but always only virtually through the auspices of the computer and its ethereal-like digital extensions. Will this change our idea of reality, of who we are, of what being in the world actually means? Whereas until recently we believed that these supposedly (always this big "supposedly" looms because we really do not know) projected archetypal images were issued exclusively out of our minds onto the environment, now we can transmit our images through a camera not only through cyberspace but into another's time and space holographically—albeit, in the latter case, in a limited way.

Actors in a Cyber World

Our ingenuity is setting us up to be actors on more than a cyber-space stage. If skeptics believed that gods emanated solely from the screens of our minds and that these appearances formulated our worldviews, we will go beyond just believing this and will recreate ourselves as gods through digital transmissions. We do not know whether the electronic communications we send into space are destroyed completely (remember, we need just a smidgen of any part of an holographic image to activate it in its entirety). For all we know, these signals can travel forever through space. Will our earthly transmissions eventually be received by intelligent life on some distant planet? SETI (Search for Extraterrestial Intelligence) continues to systematically listen for signals from space, as their scientists work on developing even more sophisticated antennas than already exist to receive cosmic signals. SETI hopes to prove the existence of extraterrestrial intelligence, and scientists working on the project anticipate sending and receiving better and clearer signals, eventually from satellites that are launched into space. Although we on earth learned about radio waves when they were discovered in the late nineteenth century by Italian Marchese Gug-lielmo Marconi, an electrical engineer, radio waves have always bombarded earth from the stars, other galaxies, and even planets such as Jupiter. This bombardment was accidentally discovered by Karl Jansky, a radio engineer, in 1930, allowing him to deduce that the interferences received by shortwave radio originated from the center of our galaxy. Not surprisingly, we remain interested in elec-tromagnetic signals from the cosmos that we now receive through our antennas.

Electromagnetic biological reception has never been a consider-ation, even though a myriad of species on earth operate instinctively by receiving data directly from the sun or other cosmic sources, attesting to the fact that such biological reception of cosmic infor-mation is part of our natural world. Is it possible that our earthly technological creations, transmitted electronically, can become electromagnetic illusions that aliens on some other planet in some other solar system will be able to activate, because these aliens possess sophisticated radio receptors or receive cosmic images through the same genetic material as that which runs through our veins? Did our prophets—and do our visionaries—see their visions because a common electromagnetic cosmic currant activated them, either technologically or biologically? If this is so, then our virtual reality digital projections might well be destined to mesmerize the imaginations of aliens on another planet, who perhaps are just on the brink of attaining intelligent consciousness.[144] Would these aliens assume, as they evolved in consciousness, that what they saw projected in their time and space (in one of the assorted technological vehicles that we produced on earth) was just a product of their fantasy? Would they think of us as aliens? Would our technological accomplishments appear as UFOs to them? Would they think them to be vehicles of the gods? Would they have anything to say in their mythologies about our warring mentalities?

We do not have to look very far afield in our literature to see descriptions of what we currently label as being just imaginary wars. Joseph Smith's account in the Book of Mormon uncannily describes what appears to be a nuclear explosion.[145] How did he conjure up a story of what so closely describes a nuclear explosion? How did Smith know, in the nineteenth century, what the results of a nuclear explosion would look like? How could a biblical author imagine an explosion that today reads like it might have been a nuclear one, as when God rains down brimstone and fire out of the heavens and the wicked cities of Sodom and Gomorrah go up in smoke?[146]

Because of our technological prowess, we appear poised to confront the mystery that underlies the phenomenon of appearances that have been reported since recorded history. More people are confessing to seeing appearances; more people report physical contact with aliens, of visiting their spaceships; and of course, more people confess to their out-of-body experiences (OBEs). Physicists are studying the possibility that parallel worlds exist, and they hypothesize that in each world we simultaneously take on different roles because of our parallel selves. Each world can be activated and lived in and called into life through a quantum observation. Quantum dynamics allows us to imagine many possibilities but to choose only one that then is activated as real. These dynamics allow us to reflect on whether we can simultaneously coexist in many worlds at the same time, if only we learn the trick required to do so. Deepak Chopra does more than just tell a pretty story in his book *The Return of Merlin*.[147] He describes parallel worlds in action and provokes us into thinking about whether simultaneous virtual dynamics might actually underlie spacetime. Hence, his book evokes a far deeper message than one of transgressions into mythological interactive time. It provides food for the imagination and sets a goal for the evolutionary brain to strive toward.

When we ponder the direction in which our cyber world is taking us, one that involves holographic and 3-D projections through space, and when we reflect upon the cyber stages being set up for us on earth, we have to wonder whether all this intra-earthly crossing of spatial and temporal boundaries is happening to prime our imaginations for a reason. Whereas before the digital age, we did not have enough scientific evidence to uncover the underlying reasons for our projections, we now have enough evidence to change our perspectives of how we have created our worldviews.

No longer can we assume that projections emanate only from our visual brains onto the screen of our natural environment in a unidirectional flow. Instead, our technological expertise is revealing to us that there is more to our visual projections than we have come to believe; that something else may be happening that involves more than just our projection capabilities; that we may be participating in an interactive biological cosmic energy that is always present and has always been present. The dynamic that Mircea Eliade posited in his work *The Sacred and the Profane*,[148] that primitive human beings felt a strong need to worship in a sacred space where a sacred energy could act upon them and be received by them in a special way, may indeed be a legitimate idea, because this felt experience of the sacred is still happening today to a majority of people in the world. Worshippers in a mosque, church, or synagogue need these specially created, often resplendent pala-tial spaces in which to worship because these are the places where they *feel* they receive sacred energy.

Undoubtedly, we have a very strong desire to relate to some-thing above in the heavens, something divine and otherworldly that we experience viscerally here on earth. Religious people believe fervently in this otherworld because their religious founders gave witness to it, while others believe just as fervently because of extant reports by abductees and other visionaries who claim to have seen appearances of aliens or gods. Much like Purves, who investigates and provides proof of inherited visual reflexes, I suggest that we have archetypal instincts programmed into our genes, a part of which harbor this strong desire to reunite to the cosmic source from which we originated. This is clearly evident in what our tech-nological instincts have led us to accomplish so far. All this, from cosmic inspired technology , has been fueled by our belief systems from the beginning of recorded time—and for very good reasons: cosmic evolution needs to take place too.

Beginning with our visual systems, we have considered the innovative steps toward artificially creating what looks like mate-rial reality. As already referred to, the idea of a holographic image traveling through cyberspace is not that far off the mark—the vir-tual reality image of the homunculus in the machine might some-day be projected into the real-time space of the environments of alien biological beings somewhere else in the universe on other planets. In such holographically projected travel, when mutually agreed to, either we or the alien would have the choice of deciding in which space the visit would materialize. How would this happen? Obviously, from today's scientific perspective a person could never materialize as a real body but only as a virtual representation—yet one so real it could not be distinguished, even by touch from the actual person, who would, of course, be in another place. Impossible? Remember what inventor Susan Kasen Summer said about her work: "The images appear so real that they defy the brain's ability to perceive them as virtual." Would this, then, be the type of image that aliens on some distant planet could project to us? Would it be somewhat similar to the UFO images that people have seen throughout the ages and about which many fairy tales about flying boats or vehicles have been written?[149] Would the virtual projection of a human being into another's spacetime be accomplished in the same way that holographs are now created—through similar yet unknown means that were more than simply "optical illusions"?

Optical illusions are part of the language of the cosmos, too, as cosmologists discovered when they analyzed data entered into their computers about the Great Wall of galaxies that is spread across the universe in a span of half a billion light-years. They found that there was no symmetry to the wall at all; rather, galaxies appeared clustered on surfaces of soap-like bubbles, apparently to avoid the holes beneath them. This Great Wall is impenetrable, meaning that we cannot see through it. It breaks down ideas of how long it took for the universe to form voids. Even the cold dark-matter theory, which would account for the voids by helping to slow down the evolution of the universe and hence allow more time for voids to form, fails. The Great Wall sets itself up as an illusion, as though it were there to block any further ideas of the symmetrically expanding cosmos that we might have.[150]

The manifestations that people see, which they swear to be real appearances, are phenomena that we have never been able to explain—and there have been all kinds of attempts to explain them.[151] It now appears that we are being forced to reconsider the dynamics in which imagination and vision engage, as we begin to close the gap between what we thought were simply our imaginary abilities to project appearances into real-time and space and our actual abilities to create holographic projections in them. The cosmic myths on which we have been weaned are not only part of religious language but also part of scientists' desires to become astronauts— to risk their lives flying into the heavens in great "boats," and so forth. It would appear that we ourselves have projected or received these appearances in order that we could, at this point in our evolutionary history, actually produce spaceships and become god-like, like our appearances.

Are these archetypal cosmic images trapped in our DNA, while at the same time floating free in an electromagnetic cosmos? Are cosmic appearances necessary because the cosmos is a self-referencing system, perhaps even an *optical illusion* in which human brains take part during a cosmic evolution?

CHAPTER 5

SPACETIME BIOLOGIZED

Before we consider the dynamics of the phenomenon of breakthrough appearances in more depth, we need to wend our way
through a few biological and physical threads that have to do with
Einstein's notion of "viscosity," a notion of spacetime bound to
and dragged around by earth and other planets. This notion set me
thinking about whether we could, for example, postulate a dynamic
that actually weaves a "fabric" (physicists' word for it), such as
that proposed by Einstein's spacetime, throughout the cosmos as
the cosmos expands. Could we think of the result of this weaving
as a biological product that contains life and that permeates the
entire cosmos—what Christian de Duve believed to be "a biological imperative"?[152] Although I do not pretend to attempt such an
exercise here, I will substitute Einstein's mathematical model of
spacetime for a biological model of spacetime in order to initiate a
change in our *imaginative perspective*. This does not mean that my
proposal is all that new—philosophers have always discussed the

nature of spirit—nor does it mean that all of our physical models need to be discarded; rather, they need to be preserved and incorporated into new ones. Ostensibly, the cosmos is permeated by electromagnetic energy, and so are we. Our brains operate through the electrical signaling of neutrons that, in the end, organize our thoughts and our bodies. So, instead of thinking about the cosmos through the lenses of a geometrically biased notion of spacetime, I suggest we rethink it and see it as an organic one.[153]

Presently, we have so many scientific mythologies about what constitutes space and time that hypothesizing about them in an organic way seems refreshing. The universe, as David Hume reminded us in his *Treatise of Human Nature*, was thought of by the ancients to be a living organism. More recently, J. E. Lovelock's Gaia theory[154] was imagined along the same vein—that everything in the universe throbs with organic life. What is different about the hypothesis presented here is that we now have a way of supporting it, as we have biological evidence that biochemical life, at some level, actually does exist in other parts of the cosmos. Still, even the staunchest believers in cosmic organic life have not actually accounted for how they would conceive of space and time functioning organically.

The Integrity of Biological Spacetime

So let us assume a biological dynamic that permeates the cosmos that includes an Einsteinian notion of spacetime. (This is a necessary postulate, even though I cannot begin to guess, nor can physicists, at the substance of its materiality.) Unless we postulate that a materiality of this nature does exist, however, we cannot account for the phenomenon of appearances in spacetime, just as we cannot account for them in the mathematical arena of spacetime that we presently postulate. In thinking organically about spacetime, we can more easily respond to notions of genetic compatibility and holographic crossovers into each other's spacetime, because all spacetime fabrics would be constituted out of the same biochemical

"stuff." Since the organic substance of spacetime would be receptive to crossovers because of its biological nature, it is not difficult
to postulate why spacetime appearances can transcend the known
laws of nature to which we are presently bound. But we are prolep-
tically ahead of ourselves.

Einstein, as well as proposing that spacetime might be dragged
about by each planet independently, also proposed that light needs
to travel through space in order that we measure the curvature of
spacetime. Scientists worked for years on perfecting a gyroscope
in order to *prove* that Einstein's hypothesis of the fabric of spacetime was true.[155] In 2007, after forty-nine years of the starting
and restarting of the project, scientists finally released data that
confirmed a theory in general relativity—that a rotating body in
space would pull the "essence" (their word) of spacetime around
with it.[156] Thus, Einstein was right: our planet is twisting the fabric of spacetime (known as "frame-dragging") around with it as
it spins. Scientists undertook this experiment because it was vital
for the future of physics. If it had failed to prove Einstein right, it
would have falsified a great deal of the physics presently believed
to be true and no doubt changed the course of physics as we know
it today.

Because Einstein was right, and spacetime is dragged around
by our planet like molasses (imagine the earth immersed in a viscous fluid, as scientists do), will this now change our notion of spacetime in any way? An Einsteinian understanding of spacetime,
however molasses-like, is still a very abstract notion. This abstract
notion of spacetime makes up our present understanding of the
dynamics inherent in the universe, and we are more or less intimi-
dated into believing that there can be no "otherworldy" spacetime
intrusions into it. Now that spacetime is proved to be a dragged-
around fabric and confirmed by scientists to be a viscous fluid,
they will be given a licence, of sorts, to rethink the whole idea of
spacetime. They could, if they wished, change their notion of it
from an abstract physical model to a biological one. We, however,

unbound by physicists' notions, are free to re-mythologize space-time in ways that suit our imaginations and cosmic views more auspiciously. Therefore, I present the following in the spirit of theorization, as physicists and cosmologists do.[157]

Because scientists have proved that the spinning earth does, in fact, create its own fabric of spacetime geometry, a plausible next step could be to show whether our spacetime geometry is different from other planets and their spacetimes. If they did this, they would make a dent, scientifically, into the question of what spacetime *biology* consists of. Until research of this nature is available, I will assume, for the purpose of my thesis, that these goals have already been achieved and that the biological fabric of the cosmos—and hence, the spacetimes—are compatible with each other.

So what is the greater purpose of this chapter, which thus far is based on suppositions? It is to introduce a laser-like cosmic interference into our model of biological spacetime. This will provide us with a way to demonstrate that apparitions can be produced by the cosmos in more or less the same way as our inventors on earth produce holographic apparitions. The cosmic biological paradigm that I am proposing will not exclude the previous mathematically derived paradigms, but it will exclude *mathematical measurement*. It will incorporate quantum dynamics in order to explain the dynamics of these kinds of spacetime intrusions. Here, for example, the work of Hugh Everett should be recalled, because he was the first to posit a multi-universe capable of the simultaneous superpositioning of worlds. Instead of a detached observer's measuring and thereby collapsing the wave function's possibilities into one outcome, Everett did two things differently—he allowed the microscopic and macroscopic worlds to coexist simultaneously by introducing a universal wave function; and he immersed the observer into the system. Instead of a wave-function collapse into one object or one world, a bifurcation occurred, thereby allowing for the existence of an observer in many worlds simultaneously. In Everett's theory, many existences and therefore realities were possible. Despite his pioneering contrarian views, however, he still was well mired in the theoretical world of equations and mathematics.[158]

Aside from noting that the idea of bifurcating wave functions has been part of the imagination for some time, we also need a venue for light that would play an important role in these spacetimes' dragged-around essences, if they are to interfere with each other and produce holographs. This is important if we are to explain apparitions as holographic interferences from other spacetimes. And indeed, light does play a significant role in creating cosmic mirages through gravitational lensing by splitting a single object into two or more images—the biggest single illusion is perhaps the Great Wall of galaxies. If, instead of a double-object mirage (as scientists think of it presently), it was thought of as a hologram, we would have the following scenario: a cosmic hologram would be produced by an interference of the pattern of light with itself in the same way that a holographic image is produced on earth. In thinking of these cosmic mirages as holographic in nature, we can better understand the dynamics of a universe that manifests its past, present, and possibly future simultaneously. Scientists already suspect that what we are measuring as light from distant parts of the universe is creating illusionary images, caused by light interfering with light.[159] It is Einstein who identified this interference as a gravitational lens that distorts our perceptions of spacetime.

So, we have dared to unravel the space and time that Einstein embedded into our imaginations by adding to it an additional dynamic, based on a bifurcating model of light on cosmic DNA and on biological spacetime. This is not to say that Einstein's brilliant insights into the curvature of spacetime and the relativity of time will ever be discarded in favor of a new theory (although there are physicists who are challenging Einstein, like João Maguerijo, who is proposing a new theory of a varying speed of light).[160] It is to say, however, that Einstein's theory would become even more important from a biological perspective of spacetime than it has been as an abstract theoretical model.

I am not the only one thinking about spacetime in a new way. Physicists in the past have introduced theories about the essence of spacetime. Gerhard Staguhn,[161] a journalist and science writer, writes cogently about physics and cosmology, including the concepts of Einstein, Planck, Bohr, pointing out the contradictions that exist in their theoretical guesses about what the cosmos is or is not. In his conclusion, Staguhn cites French nuclear physicist Jean E. Charon's thesis that embraces a spiritual physics with which he (Staguhn) takes exception. Charon postulates a special kind of spacetime structure in which electrons are thought to be "matter particles endowed with a kind of elementary consciousness" or as Staguhn refers to them, "spiriticles." Charon believes "these electrons were closed microcosms with a space-time structure that no longer obeys the known natural laws. They can store information of their own 'history,' the sum total of all interactions with other particles since the beginning of their existence."[162]

While Staguhn may have a problem with Charon's theory, since it does not include photons and neurons, I do not think the problem is insurmountable, because more and more insights into the nature of photons and other particles are being uncovered (we shall come back to this below). At this point it is important to recognize, however, that Straguhn has indeed called our attention to a good idea by labeling Charon's concept "spiriticles," because they are, in my view, quite compatible with biological substance. Straguhn points out that in quantum mechanics, when physical and metaphysical are reduced to infinity, they become one idea. Yet Straguhn's physical idea remains essentially in the realm of the invisible material of electrons, protons, photons, quarks, and so on. He never interprets them as biological in their physicality, even though he states that physicists are apt to suggest that a primary force acts in everything,[163] which would imply that we are the products of the very first "atomic" moment of creation. Still, Straguhn's astute

analysis of the state of physics and cosmology and their close relationship to metaphysics, like Talbot's, also ends on a Taoistic note,
which points away from the biological toward the abstract. (Funnily enough, although Eastern religions take most of their koanistic
examples from nature, the koans, nonetheless, remain materially
difficult to grasp and so fall easily into the abstract category.)

Must we think, however, in terms of a spacetime structure that
obeys only the natural mathematical laws on which we have been
weaned, or can we extend our imaginations beyond these constrict-
ing laws? The fact is that our visual systems are capable of showing
us anomalies that intrude into the Einsteinian notion of spacetime
and challenge us to contemplate the breakthrough appearances
that we have been considering against them. Why do some objects/
subjects manifest themselves to us as though they come from
another spacetime? What is going on here? There is no debate on
this subject because, as already noted, scientists insist only on rig-
orous scientific proof in respect to such appearances, and they continue to challenge those who are not scientists and claim to have
seen otherworldly appearances to provide hard evidence as proof.
Mainly, it is thought by most rational-minded or left-hemisphere-
oriented academics that such appearances are what Kant claimed—
simply subjective mental phenomena. And this is, indeed, what
they are from the point of view of human biology. But I am convinced that aside from human biology, the entire scenario of apparitions can be played out cosmically as well, because the cosmos,
too, is biologically structured.

If this is so, what in consciousness generates these phenomena
so that in the end, they come dressed in similar universal archetypal biological clothing as our own? Physicist Charon's thoughts
focused on electrons. He stayed within these parameters—as far
as he was concerned, his immortality was connected to the primal electronic stuff of the universe. Staguhn, on the other hand,
questions this kind of thinking: "What is the use of a mind that is
split up into innumerable electronic mind-particles?" For Staguhn,

"electronic immortality has little consolation to offer."[164] We, on the other hand, must keep Charon's thesis—his "spiriticles"—in mind because, as we will encounter, these "immortal" electrons have much to do—*but not everything to do*—with activating in consciousness the apparitions and visions we have been considering.

We continue, then, with a biologized space and time instead of staying with the mathematical/geometrical/Einsteinian paradigm. Rather than think in terms of a time machine, traveling back into the past or forward into the future in some Wellsian fashion, I propose that we locate space and time in our own genomes. Wells' ideas were bound to the mathematical and linear ideas of space-time; ours will be bound to the biological. As clever and imaginative as Wells was with his futuristic scenarios, time machines were never invented except in movies, nor will they be in future epochs. Organic spacetime, however, leaves us with a completely unique model for transcending space and time, which makes even more sense than the time machines that science fiction writers—and even professional scientists—propose as being possible.[165] Before we discuss how organic spacetimes break through into each other, we need to step back and consider how a biological notion of spacetime might work.

Superpositioning Genes

A pioneer cytologist, Barbara McClintock, speculated how chromosomal dynamics worked well before Watson and Crick discovered the double-helix structure of the DNA molecule. Without going into details of the genetics of mitosis and meiosis, I will get to my point more directly. McClintock's research was mainly with maize, which allowed her to study genes and their mutation rates in an accelerated way. Essentially, she based her findings on the results she saw in the fields and what she saw occurring naturally or even when manipulated. This led her to very different conclusions from those of her colleagues. While not entirely spurned, her revolutionary ideas, which included an idea of genes turning on or off, were not accepted. Classical genetics required a static model— beads of genes strung in order on the chromosome, just as Einstein requires a constant speed for light. McClintock, on the other hand, saw something quite different going on in her maize genes. These were not static beads on a string in a chromosome but quite lively and involved in what she called transpositioning—and this is precisely what I propose for spacetime dynamics.

With the discovery of *molecular genetics,* the classical static genetic models were called into question. Young scientists in the field came up with startling conclusions. Using different techniques and their own methods of research, they affirmed McClintock's theories of transpositioning. Suddenly, "jumping genes" in chromosomes were normal vocabulary in the field of genetics, with all kinds of chromosomal data, sometimes in chunks, being identified as transposing itself within the genome. Evelyn Fox Keller, who wrote a biography about McClintock, points out that to do justice to McClintock's vision, we would have to provide for "a concept of genetic variation that is neither random nor purposive—and an understanding of evolution that transcends both Lamarck and Darwin."[166]

Today, our theories about genes have evolved much farther than McClintock's radical ones. Geneticist Fred Gage and his colleagues at the University of Michigan have posited that "jumping genes" probably alter the wiring in our brains, creating our unique personalities—the reason why twins, although otherwise identical, do not think exactly alike. Medical geneticists believe that jumping genes, or transposons, as they call bits of DNA today, move around freely in our genome and can rearrange our mental structure. They can create havoc by traveling to new places or by pasting copies of themselves in random places on the DNA. With this new discovery of transposon LINE-1 (humans have a much higher proportion of in their genomes than other animals), the question as to whether or not this is going to have any effect on learning, memory, or behavior remains to be investigated.[167]

With the shuffling around of genetic material now clearly identified by scientists as occurring in the genome and particularly in the brain, where LINE-1 seems predisposed to alter genes for unique brain functioning, it is not that great a leap for me to suggest that our genomes contain archetypal genetic data on particular chromosomes that is perhaps more easily reshuffled in some people than in others. (For example, the discovery that what was thought to be a strictly recessive, extra copy, turned-off X chromosome in women, is actually 15 percent active.[168]) Scientists are also rewriting the laws of heredity by decoding what causes genes to turn on or off. The field is called epigenetics, and it studies the genetic switching system—which genes to turn on and which to turn off to fight cancer, for example. Epigeneticists have determined that although most genes do not go this route, there are genes that can be passed over to future generations. These passed-on genes do not produce DNA mutations but can make us susceptible to something like bi-polar disorder. Epigeneticists are looking for patterns that could reveal why certain biochemicals are released in some people but not in others.[169] Thought of as a second genetic code these traffic genes are like dimming switches in our bodies, controlling how much or how little of certain chemicals are released—or not released.

There are at least thirty thousand genes that carry instructions for coding proteins to keep our bodies going. Since the visual brain could be involved with a turned-on gene receptive to certain chemicals, it is possible that perhaps not only an individual's jumping genes but those of a past generation's that he carries are responsible for producing the chemical space where hallucinatory images can arise. Dreams, of course, reveal archetypal material readily, perhaps because different chemicals are released in our brains during sleep, while consciously projected archetypal material, such as appearances, may be more prone to occur in certain conscious states than in others. The fact that elderly people, with no mental problems but with visual problems, hallucinate images is one clue about the reality of visual projections. The other group are our schizophrenics and people with bi-polar disorder, who have mental problems but no visual problems. Schizophrenics hallucinate and hear voices, caused by what is generally thought to be a chemical imbalance in their brains.

Aside from the chemical aspects that might affect the visual brain, there are electromagnetic aspects that probably act in tandem or even produce the chemical reaction going on. We know that genetic traits caused by mutations over time are often specific to certain populations. Such a variant gene is referred to as an allele and is found, like a gene, on a particular chromosome. But suppose both such traits are turned on at the same time—the original and the variant. The person would then have two traits that are programmed to do the same thing.[170]

Let us take vision as an example. If a person had two alleles to process electromagnetic light as it enters the eye, it might be possible for the two to act simultaneously, and this could superimpose incoming electromagnetic fields. Such a person, born with two turned-on traits, both of which if acting as receivers for specific frequencies of light, might be able to process normal environmental 3-D space, while at the same time being visually capable of superimposing holographic appearances from the cosmos upon it. Or, elaborating on this from a slightly different perspective, a person who might have inherited a double dose of a "spatial frequency trait," both of which are turned on, might be visually predisposed to accessing his ancestors' experiences that are stored in his genome's database, and he would thus be able to juxtapose both ancient archetypal imagery with real-time imagery onto a normal 3-D spatial environment. In view of the evidence we have amassed on the subject of genes that are turned on or off and the effects of a double dose of turned on alleles, I think hypothesizing like this about visual traits is not out of line, especially in light of our conclusions in the previous chapter. The very thought that electromagnetic signals might enter the brain in multifaceted ways, depending on the genetic traits that the individual inherited, leaves room for many other questions, such as, do so-called "demon-possessed" people and schizophrenics suffer from a visual transpositioning of genetic material within the genome that, as Gage suggests, wreaks havoc with their personality?

But let us concentrate for a while on the electromagnetic signals that we have been discussing. They are said to be a hundred times faster than chemical signals and are responsible for energizing every cell in our entire body, from DNA to the very important microtubules, to the all-pervasive bodily connective tissues.[171] There is no question here that we earthlings are biologically constructed to be electromagnetic receptors; lightning, for example, kills seventy-three or more people a year in the United States, far more than are killed in tornados and hurricanes. Scientists are attempting to learn more about the causes of lightning. Right now, it is thought to originate in the cosmos. Lightning-struck people say their personalities are changed forever, and research based on fMRI scans confirms this.[172] Doubtlessly, the mystery of consciousness lies buried in the interaction between the biochemical and the electromagnetic interactions in our brains, particularly our visual brains.

The suppositions of electromagnetically produced cosmic visions appearing in front of our eyes are not idle speculations when one considers that technologically today, we are able to beam images directly onto the retina through what is being called "laser vision." The U.S. firm Microvision developed a system some years ago that can project images on top of a normal field of vision. This means that an individual can continue to work on whatever she is doing and see any information she needs floating in front of her, not farther than arm's length away. These display systems, which use a scanned-beamed technology (lenses, mirrors, lasers) and electronics to process signals from data or image sources, are already in use. If, technologically, we can achieve superpositioning of desired images directly onto our retina, surely human beings, as well as the cosmos, are biologically capable, with electromagnetism on their side, of doing the same thing naturally. Consider the fact that during photosynthesis, the energy of a single proton can be in many different states at one time during an interaction between chlorophyll molecules in a bacterium and the sun's energy. Scientists, as Tim Folger has reported, think of this interaction as a weird quantum mechanical happening and refer to it as a "superpositioning."[173]

The value of positing a biological essence of spacetime found in a yet-to-be-identified gene or genes would lie in the resulting ability of the visual brain to superimpose images that it receives. A biological dynamic involving jumping spacetimes clearly allows for imagining a superpositioning of appearances of one spacetime upon another's because of attracting, compatible, crossover spatial cosmic essences that are programmed to be receptive in the genome. Superpositioning of spacetimes, by certain people, because of biological spacetime's compatible spatial essences, means that we now have a way to account for the breakthrough phenomenon of apparitions of alien travelers or spacemen, which have been reported throughout the ages by peoples all over the world. I will not dwell on the variety of apparitions that have been amassed, for we have covered enough of these stories in the first chapters of this book to aid the reader in determining what is going on in these otherworldly sightings. I simply invite readers to open their minds to paradigms other than the mathematical/physical model we are trapped in today and to think of these supposedly folkloric phenomena in the light of a new thesis—a biological model of spacetime, which, as I have been arguing, will include cosmic as well as human aspects.

Pop-In Spacetimes

To ground the biological spacetime dynamic that I am proposing we will examine a few physicists' ideas of spacetime. Their imaginings are legendary and leave dangling to be verified or falsified all manner of universes and the dynamics that go with them. Physicists, as part of their fertile imaginings, imagine the existence of virtual particles in vacuum states because, they say, outer space could never tolerate a vacuum—that is, space can never be completely empty. Rather, there must always be some residue, and that residue must be accounted for by the notion that empty space can never be really empty but must be filled with tiny particles that are always popping into and out of existence. Tufts University's Alexander Vilenkin has extended this paradigm to make a further analogy. He posits that if this is true, why not have universes popping in and out of existence?[174] Likewise, we can take his lead and instead of theoretical universes, posit biological spacetimes as popping in and out of other spacetimes. Our theory can be substantiated to a small degree by "jumping genes." A biological paradigm of the reality of the multi-possibilities of spacetimes in genomes, as discussed above, is possible and verifiable if we concede that our visionaries can, indeed, manifest simultaneous images of other spacetimes onto their own real-time space, by virtue of the genetic mutation they might carry. There is a curious interscientific reinforcement going on in these ideas of pop-up spacetimes and jumping genes, hence the reason I can propose that this may, in fact, be what is happening in our reported real-time experiences of apparitions.

Phenomena that pop in or break through into our real-time spacetime by a superpositioning-interfering event would create a co-mingling and suspension of two different spacetime essences in one transparent vision of spacetime—a co-mingling that is usually reported (except by abductees) to be of very brief duration and/ or as having taken place in an unreal sense of time. It may not, however, always be that this superpositioning of spacetimes happens accidentally when certain conditions prevail. It could be that a particular spacetime's genetic trait in a person is more susceptible than another's in allowing such penetrative co-mingling. Undoubtedly, our present-day physical models limit our imaginations. This is the reason we need to free our minds and transcend our presently dominating spacetime paradigms. Alexander Vilenkin, for example, speculates that some universes are more fecund than others and hence might be teeming with "millions of inhabited solar systems."[175]

When a spacetime interference occurs, at least one spacetime (the alien's) becomes transparently superimposed onto the seer's. From the seer's point of view, she might feel as though she had transcended her normal mundane state of consciousness. Abductees constantly report feelings of having transcended the spacetime that they know on earth. It could be that otherworldly apparitions appear to certain people simply because something about the visual brain triggers the breakthrough into a spacetime that allows them to see these otherworldly appearances clearly, such as we reported earlier, which happened to Goethe on his horse-drawn journey to another town.[176] For example, the apparition could be triggered by a sudden burst of electromagnetic frequency acting upon the genetically vulnerable person. We should cite, too, people's reports of space travelers and their saucer-like airborne vehicles, which appeared to simple country people both in the United States and in Europe in the nineteenth and early twentieth centuries, even before such vehicles were invented. This suggests a cosmically induced dynamic or information that has been stored in our genome's database that is retrievable.

It ought to be noted at this point that unlike what quantum reality physicists posit—where the observer's consciousness produces the material event and stops all other possible realities from "becoming"—the apparitional dynamic that I am proposing ends up hovering in a transpositional spacetime; it does not become material, even though it appears as though it is. Since quantum teleportation was discovered in 1993, it has been aggressively studied with some tangible results. Although there are mass sightings, most people who experience such an event do not see a tangible appearance that others who are with them may be seeing. The appearance has appeared solely to them because of their genome's spacetime receptive alleles, for reasons unknown to the seer. Also, not all interferences are peaceful incursions into a person's spacetime. Abductees, like schizophrenics, appear to participate more in a "destructive interference" than in a "constructive one." Often, judging from abductee reports, it is as though the interfering spacetime apparitions control the scenario—that aliens from the interfering spacetime can override personal consciousness and take control of abductees' consciousness. We can only speculate that perhaps something like this alien intrusion scenario has occurred before in time and is programmed in our DNA.

Other types of phenomena can be categorized with those just discussed above—for example, the Bible's reference to angels appearing to people; to the resurrected Christ's appearing to his disciples; and to the Virgin Mary's appearing to saints. These types of apparitions seem also to take place as a result of a visual system that purposefully and consciously superimposes a holographic projection onto the environment in real-time. I mean that this presents in the same way as the visual brain displays a three-dimensional image when one looks at a stereogram specifically produced to create such an effect. The nature of these apparitions do not suggest that they are projected at us from the cosmos and another's spacetime, since usually only one person (or sometimes a small group, perhaps influenced by this person) is privy to them. In fact, everything about this genre of visions, often seen by children and female adults (and usually containing rather dreadful repetitive messages that condemn the inhabitants of earth to fire and brimstone unless they adhere to the apparition's God's wishes), indicates that a biologically related dynamic is going on, albeit conditioned in content by a particular time's culture.

One thing that we ought to keep in mind in respect to apparitional experiences is that the moment we try to explain them, through one so-called "simple" universal spacetime law, we close all other venues, limiting our imaginations and thus our considerations of these apparitions to a particular confining paradigm. The one we use at the moment limits us to explaining apparitions through a paradigm based on physics—an impossible task, as like oil and water, apparitions and logic do not mix. Ultimately, what we are dealing with in apparitions is an interference of unconscious genomically produced images consciously experienced—a visually based experience that transcends geometrically and mathematically (conceptually) based ideas of what can occur in spacetime. Essentially, we still live by the rules of the Einsteinian model of four-dimensional spacetime. Everything that happens to us must be explained through and by this model of spacetime, including appearances. Yet increasingly, it becomes apparent that these other unexplainable moments of spacetime, which intrude upon our own and which have done so for so many centuries, must be generated by an entirely other genus of spacetime. It must be a spacetime that cannot have anything at all to do with abstract spacetime geometry, because an interfering biological spacetime, whether caused cosmically or genomically (actually, an inseparable juxtaposition), breaks all the rules governing our present paradigm.

CHAPTER 6

SCIENCE FICTION, SCIENCE, AND EVOLVING IMAGINATION

Star Trek

Science fiction has worked hard, imaginatively, in its attempts to break through the limits set by the speed of traveling light, but light, along with the human body's present inability to cope with a lack of gravity for long periods of time, still remains the stumbling block for physical space travel. Scientists are beginning to concede that unless we circumvent the physical limits of time travel, we are more or less stuck on earth and/or its immediate environs, which is not to say that this limited kind of space travel cannot exist (for example, we have successfully journeyed to the moon) or will not exist in the future,–our astronauts still dream of journeys to Mars. Einstein has provided us with an understanding of space travel that is entangled with the relativity of time. The faster a space-ship would carry earthlings away from the earth, the slower time would go for them. Coming back to earth would mean that they would find their beloved families either very old or dead and a new generation born—or possibly many new generations born—during their absence. This kind of space travel, based on the limitations of the speed of light, we are direly aware, will not meet our desire to see what is out there in the rest of the universe. It is also the reason why science fiction's "physicists" have been positing other kinds of paradigms for space travel that are not based on actually getting into a spaceship but have to do with being transported through spacetime in another fashion altogether.

To my knowledge, the notion of traveling through spacetime, due to the superpositioning of spacetimes as biological occurrence, has not been broached before. However, we do see something like this happening in *Star Trek*. The holodeck serves as a place to go and relax; to call up memories of home, of food, of pleasant encounters in the past that one is seeking to relive—all of which implies a superpositioning of space and time through a holographic transmission. The *Star Trek* series contains, for example, the imaginative notions of warp drive, where the spaceship travels toward an always-collapsing spacetime in the wake of a grossly expanded one. There is the "beam me up (or down)" transporter technology to move a body from spaceship to planet, which, in effect, overrides conventional Einsteinian dynamics. Still, *Star Trek* is essentially the product of a mathematically influenced imagination, even though warped spacetime travel occurs in it and all kinds of strange transuniversal encounters with other planetary beings take place. Despite all this heady holographic-type stuff, the creators of the series still conceive of space travel as needing a spaceship to transport its crew across spacetime. Although transporting bodies in a holographic pseudo-biological manner appears to be happening in *Star Trek*, space travel itself is not presented in this way—if it were, then there would be no action-packed story to tell; no high drama to unfold about spaceships at war in space.

Should we pay attention to these science fiction schemes? Well, physicists like Lawrence M. Krauss do.[177] For the most part, Krauss concludes that the *Star Trek* producers have a lot of the physics right (although not all) in their construction of this space fantasy, which brings us to the oft-repeated question of whether science imitates imagination or vice versa. In the case of *Star Trek*, we have to ask to what degree this television series has formulated our present views of spacetime and spacetime travel—and if it has, what does this say about our imaginations? If the answer is that imagination imitates science, then our imaginations are surely leading the way. Increasingly, it looks as though this might be true, as our evolving imaginations still continue to look to the heavens for deep-rooted answers for the absolute reason underlying the existence of the cosmos. Unbeknownst to scientists, they are not so much plowing into uncharted cosmic territory as they are actually digging deep into their own biology and tapping into the essence of spatiality *within themselves* for answers to their cosmic puzzle.

Historically, we note that imagination evolves, generally, only in small incremental ways, one step at a time. For example, a television ad reminds us that airplane travel is passé, with the implied inference that trans-earthly *holographic* conferencing is in. Subtly, an ad like this opens the way to affirming the possibility for holographic communication; for instance, holographic teachers are projected to students in remote schools in Britain as part of a teaching experiment.[178] Even though human evolution is perceived as taking place very slowly, over billions of years, we who live on earth at this time think of our own technological evolution as proceeding exponentially. Although technology is influencing our thinking about how we can travel through space on earth, it still struggles with the larger, perplexing demands of travel between galaxies. At the moment, I believe that our minds are surreptitiously being prepared for the transcending of spacetime holographically. This is a beginning of sorts, because if we are to realize our dream of exploring the rest of the universe, we have to change our basic ideas of how to transcend our present mathematical notions of spacetime.

In the previous chapter, I began by suggesting that we think of the cosmos as organic in substance, with the fabric of spacetime a connective tissue laced throughout it. Out of the cosmos' unknown dynamics, we organisms on earth began germinating some four billion years ago. That which caused life to begin remains a deep mystery that challenges science. My invitation to the reader is to imagine that the cosmos' organic substance permeates many spacetimes, which surround many planets throughout the cosmos. This idea that the universe is organic is a theory that de Duve held and one that is now being confirmed in the icy matter and rocks of comets that fall to earth or by material retrieved directly from outer space by our robotic probes. Intelligent species could well exist in many spacetimes, and I suggest that these species could be uncannily similar to—or even uncannily related to—our own, because of the compatible biochemical cosmic natures underlying both.[179]

As the biology of the cosmos is revealed to us more and more, we will come to appreciate why it is that our genomic inheritance is constantly pushing us to further explore the frontiers of space. This drive prevails whether we approach it through the biases of theology, or through what Carl Sagan referred to as *The Demon Haunted World*,[180] or through scientific minds imagining the existence of other spacetimes. For example, to borrow from scientific models, we could imagine a biological spacetime as bifurcating branches of a cosmic "tree," or as a biological paradigm taken from Steven Jay Gould's theory of simultaneous evolutions,[181] or as one of physicist Hugh Everett's many-worlds interpretations.[182] The kind of bifurcating model of biological spacetime that I am envisioning would have to be approached through light waves and the quantum world of waves and particles, through the world of electromagnetics, and through what we know of gravity. We cannot ignore the phenomena that acts upon our visual natures, whether they are revealed through microscopic cosmic particles or macroscopic otherworldly creatures seen on earth as apparitions.

Apparitions and Their Holographic Natures

If something like organic spacetime is true, then these break-through visions into our own spacetime make sense. They confirm that superpositioning does occur, presenting itself mainly as something acting upon us from above, and that the visual system manifests these interfering images as holographic in nature. Because the abductees report walking through walls and riding a light beam into an alien's spacetime, we can deduce that our visual system can manifest the rest of its body as a holographic body, capable of walking through walls with the abductors and taking rides up a beam of light with them. This is reminiscent of virtual reality software programs that creatively place the participant into a 3-D scene such as those used experimentally by the medical community.[183] In the case of the abductee experience, there is always a sense of a loss of earthly time upon their return, and sometimes the person is reported as missing during this time by family members.[184]

There are two claims going on here: one, that apparitions are produced by activating susceptible genes in some people; and two, that the cosmos itself superimposes these planetary spacetime images upon receptive earthlings by interfering with our planet's spacetime's frequency waves. What is common to both claims is that a spacetime's breakthrough apparition is holographic in nature, whether it is visually projected by the individual or projected by the cosmos at the individual. Ostensibly, holographic-like phenomena are reported by seers when they describe their visions or interactions with aliens that they perceive as descending to earth. During these experiences, both parties appear to transcend environmental materiality—this is most probably the reason why no physical evidence ever remains, or that which is claimed to be physical evidence can never be substantiated as caused by or left by aliens.

What I am suggesting is that this cosmic kind of superimposing of spacetimes would travel electromagnetically via crossovers and interferences with each other's cosmic wavelengths. Superimposing of spacetimes could occur not only to us but also to other cosmic beings who may or may not be more advanced than we are and who also cannot physically travel in a universe that is bound to the constricts of local time and infinite space. These types of visions, experienced by earthlings, usually have been understood as providing glimpses into the future for those of us on earth, and as Michael Heim suggests (see chapter 1), UFOs could be projections from the future.

As I pointed out earlier, although appearances could be happening in a parallel universe, some physicists might propose, theoretically, that the visions seen also could have happened billions of years ago or are destined to do so in the future, according to the way we choose to approach biological time. Allowing for the cosmically produced model of projection, one would have to grant that the spacetime travelers who happen into our time accidentally, because of electromagnetic spacetime interference, would be just as surprised as human beings are to find themselves face to face with aliens they consider to be from another spacetime. Given the nature of biological life and the age of the cosmos, all life everywhere in the cosmos could have developed at approximately the same time, because all life would have been subject to the same conditions in the beginning with the big-bang event. We know that it did not happen this way, because cosmologists' telescopes now show galaxies as being born all the time, inviting a myriad of new theories to be proffered at astonishing rates.[185]

While this may all sound like science fiction, promoting science fiction is not my goal. My goal is to introduce us to a biological spacetime that not only permeates our DNA but also permeates the cosmos. As such, the paradigm must, of necessity, include the nature of projections, from the experiencing of alien spacetimes projected at us to earthly ones we ourselves project into space.

Essentially, I believe that the cosmos is home to other beings similar to us and that our genomes contain the vestiges of the common cosmic matter we share with these beings. Our myths overwhelmingly point to this evidence, too, although they do not satisfy our scientists. Our cultures are inundated by religious models of up/down themes, divine/human natures, and divine sparks within us. Even science, which is mathematically and geometrically founded, plays a profound role by covertly using these very examples to enhance our imaginations. Let us look at some of these theoretical findings. They provide evidence that evolutionary ideas naturally occur in the human being through imaginative projections, whether mythic or scientific, and that they are created because of this desire to establish connections with the cosmic dimension that has given birth to humankind.

Crystalline Viscosity

To do so, we need to return to Einstein's spacetime fabric and to take this scientifically inspired paradigm a step farther and posit that the upshot of this hypothesis is that if the fabric of spacetime was viewed biologically, rather than geometrically, then quantum-type leaps, when seen through the lens of biological dynamics, could indeed account for holographic intrusions into another's spacetime.

In order for holographic images to occur in a particular spacetime, we need to introduce a crystalline structure into the fabric of spacetime to make it a holographic medium. Here again, only if light is involved will there be a conduit for the projection of a holographic image. As daunting a task as this appears to be, a beginning can be made. For instance, our computer inventors seem to be farther ahead in implementing the integration of the digital/biological/crystalline than scholars who are presently working in cross-disciplinary consciousness studies, who tend to stay within the narrow parameters of what they were trained to "see" best.

In technological inventors' work on the construction of quantum computers, protein memory computers, and holographic computers, one finds that they all incorporate into their technologies a crystalline structure and laser technology (necessary to make holograms), so that their new computers operate at lightning speed.[186] Crystals are light-splitting material par excellence, especially when it comes to working with lasers. Our bodies' connective tissues are laced with crystalline substance, including a crystalline structure in the brain and body that is composed of microtubules said to turn consciousness on or off.[187]

Aside from crystals in computers and crystals in the fabric of our bodies and brains, there are said to be crystals in the enveloping membrane of black holes (think of this membrane in biochemical terms). Also, there are thought to be crystals in the spacetime surrounding our planet and other planets, and apparently crystals cooperate in projecting the impenetrable "illusionary" Great Wall of galaxies. Can it be that only the inventors of the new generation of computers see the value of laser light when the inorganic is combined with the organic?

Our task now is to account for cosmic interference and how it produces appearances by introducing a crystalline substance to the essence of spacetime surrounding a planet, in order to allow for interferences of light from another planet and thus project holographic appearances into our own spacetime. Although there is proof now that spacetime is being dragged around by our planet, there is no proof that it is crystalline in substance. We can, however, examine some of the scientific theories that are currently proposed in respect to identifying crystalline substances in the fabric of spacetime.

Black Holes and Lattice Work

"Gravitational attraction equaling the electrostatic or magnetostatic repulsion" is posited by physicist Leonard Susskind in his studies of gravitation that have to do with saving information, instead of allowing it to dissipate and evaporate inside of a black hole. He argues "that information can be compressed into one lower dimension," a dimension he refers to as "the holographic universe."[188] His research into string theory and gravity has to do with black holes that are tractable. He believes that if black holes have electrical or magnetic charges (this is still a theory), information does not evaporate (as Stephen Hawking proposes). Over the years, some physicists claim they have discovered that the outside properties of a black hole have many physical qualities, such as electrical conductivity and viscosity, not unlike the viscous fluidity that Einstein thought constituted the spacetime pulled around by our earth.[189] The black hole membrane itself suggests some kind of crystalline structure similar to what Stuart Hameroff has found in the brain, and, true enough, we are asked by Susskind to think of the event horizon "as a crystal lattice," where information is stored and where the "number of bits predicted by string theory exactly accounts for the entropy as measured by the area of the horizon."[190]

Rather than the information evaporating into nothing and being lost forever, as Hawking imagines (although to his credit, he often changes his mind with the advent of new discoveries), some physicists—and Susskind is one of them—believe that evaporation of information ceases when gravitational attraction equals the electrostatic or magnetostatic repulsion. Susskind describes them as complementary. The complementarity produces "the remaining stable object," which is called "an extremal black hole." Whether Susskind means this "remaining stable object" to make itself visible as a hologram, I do not know, but his theory suggests this possibility very strongly.[191] Still, we cannot pass over Susskind's insights glibly, as they are synchronous with the type of holographic supposition that we are considering here, even though Susskind's insights belong to a different category of scientific investigation. These similarities in cross-disciplinary paradigms are important to note precisely because they unwittingly support metaphorical academic coincidences of discovery or, as some now refer to it, "memological" tendencies.[192] (Along with others, I believe that academics today should feel obligated to point out similarities in dynamics suggested in others' hypothetical models that have originated in another fields of study from which they may have profited.)[193]

Our approach to spacetime superpositioning and Susskind's theory of complementarity, which sustains itself through repulsion-attraction, invites us to imagine that holographic information also is not lost, because something like this theory applies to the dynamic we are positing. That gravitational attraction equals the electrostatic or magnetostatic repulsion makes a lot of sense in our paradigm, because it is one way to account for how two spacetimes could be simultaneously existent without averring to the quantum mechanical insistence that only one or another spacetime's reality can exist at the expense of the other's nonexistence. A spacetime superpositioning through a dynamic such as Susskind's repulsion/attraction scheme would allow spacetime holograms to be superimposed—to coexist unchallenged as a repulsion/attraction dynamic in one or the other's spacetime.

Susskind's theories leave us with a notion of complementarity that challenges Einstein's relativity theory. "In the special theory of relativity we find that although different observers disagree about the lengths of time and space intervals, events take place at definite space-time locations. Black hole complementarity does away with even that."[194] Suffice it to say that this suggestion of a new theory of the complementarity of time and space that occurs outside of the relativity of the time and space on which we have been weaned is very intriguing, because it describes the physics of the sort of tension to which a holographic notion of breakthrough into another spacetime might be subject. Surely what is being suggested by our biological spacetime paradigm and by Susskind is a notion of spacetime that transcends Einstein's theories of it. It remains to be seen what physical evidence Susskind will bring to his research of such a "holographic universe," where all information that gets in, stays in.

Holographic and Laser Inventions

Let us return to our discussion about how computer and holograph inventors are leading the way in providing us with clues as to what operative dynamics we ought to look for in our approach to cosmic apparitions. Cosmologists are trying to prove that natural lasers have been found in the cosmos, which allow them to see holographically into the previously unseeable core of a gaseous substance.[195] They are currently trying to perfect a practical X-ray laser that will penetrate objects and reveal these objects in a three-dimensional, holographic way. With an X-ray laser microscope, biologists, for example, could penetrate living cells; that is, reveal the insides of a living cell. "They could take holographic 3-D snapshots of structures suspended in the cell's plasma with details resolved to a billionth of a meter. They might even zoom down to the scale of molecules, pick out some bit of DNA, and find out how it orchestrates the chemistry of life."[196]

Our present knowledge of laser dynamics has been around for some thirty years. We are on the verge of making practical, useful instruments out of even more sophisticated X-ray lasers. Surely, we inventive, intelligent earthlings testify to the possibility that others in the universe might also have arrived at powerful laser technology and learned to use it to project holographic images of themselves into our planet's spacetime fabric. Or, failing that, it is arguable that the cosmos itself has found a way to utilize laser light naturally and thus project holographic images of other spacetimes when light beams cross each other spontaneously in the cosmos. I cannot claim any such occurrence as ever having happened but as cosmologists point out, a profound feature of the cosmos is that it is stratified. This makes levels of reality vulnerable to being jumbled in much the same way as different wavelengths might cross over each other in a nonlinear cosmos. Such scientific theory is based on gravity's interconnecting structures of vastly different sizes, instead of on an assumption that the cosmos' density retains an average value of one atom per cubic meter. Cosmologists who believe in the dynamics of chaos science believe that a small-scale texture of the cosmos might affect its large-scale behavior.[197] There appears to be a lot of convergence going on in the field of cosmology, as scientists posit discoveries that leave us with images of a cosmos that is vibrantly alive and constantly changing in shape and form.[198]

If one doubts that the cosmos or cosmic beings like ourselves can perform such holographic feats, consider the fact that holographer Elizabeth Downing has made a breakthrough of another order in holographic imaging by separating herself from the tradition of creating holographic visual subterfuge. Her holograms, presently created in cubes, are made of a blend of heavy metals, fluoride, and glass and display original 3-D dimensions. Unlike real objects, the images in the cube do not block light, and hence the scene looks slightly ghostly. The backyard of a house shining through its wall is the example she uses. (This holographic technique affirms the kind of scene that abductees often describe when they report seeing vehicles in backyards, shining through the walls of their homes.) "The technology is so new," Downing says, "that we're not exactly sure where the limit of visualization is."[199]

Considering our present holographic technological knowledge, it is highly probable that these otherworldly holograms that we on earth have witnessed throughout the ages—and continue to witness—as unexplainable apparitions will be accounted for in due course scientifically, particularly if our present knowledge of holographic technology is combined with dynamics that involve both quantum light and biological factors. If this kind of dynamical synthesis is achieved, it is inevitable that the future of humankind will eventually understand more clearly the universal forces involved in the projection of apparitions. We have at least five billion more years of evolution ahead of us, and even if we only survive on earth for a fraction of this time, we seem to be on track for solving more and more mysteries of the cosmos.

Holographic Instant Travel

When the prospect of laser-created interference is introduced to the biological paradigm of spacetime, as I have postulated here, and when it works through the dynamics of electromagnetic waves, quantum theory, laser light, and crystalline substance, then the possibility of holographic apparitions from other spacetimes interfering with our own spacetime begins to make more sense. Even though the laws of quantum mechanics state that an observer is supposed to collapse the quantum state of indeterminacy from wave to particle, as I have argued above, this would not occur in biologically congenial spacetimes. We know from seers that their apparitions do not collapse our spacetime or we theirs. After the interference occurs, the apparition (if, for example, it is a UFO) simply takes off and disappears at lightning speed into space. From our amassed historical reports of appearances, an alien's spacetime intruding into our spacetime holds on to its own spacetime fabric, while manifesting itself in a ghostly, superimposed, holographic animation onto our planet's spacetime. That is, it sustains its own spacetime while at the same time superpositioning itself onto ours. Consider the fact that virtual-reality artists use this technique in their work quite effectively and are able to superimpose many translucent images that are spatially distinct one from another, hence leaving viewers immersed in a multilayered, translucent, 3-D virtual reality.

If bifurcating spacetimes crossover, as genes do, and intrude upon each other like jumping genes or pop-up spacetimes do, then why is it not possible for holographic images to appear anywhere in the universe, manifested through any spacetime's crystalline fabric, when projected by cosmic laser light? If this is true, then any planet's spacetime fabric can be holographically accessed this way. Technologically, we are arriving at a point where we someday will be able to project holographic manifestations of ourselves onto another planet. We can already beam images from rovers on Mars back to earth. We have all the technology in place to send animated, speaking holographic images of ourselves into space—but on the premise that we just do not know where to send them, we send none at all when, alas, the entire cosmos is "wired" to receive them.

The biological paradigm, I propose, will prove to be the most efficient way to transcend four-dimensional spacetime in the universe and make traveling to another's spacetime a biological quantum-like event. After all, it seems reasonable to assume that a biologized cosmos can produce a holographic image that transcends the limits of light in an Einsteinian spacetime, for it will have no mass to move. It is not for naught that human beings have conceived of a "transporter" on the *Enterprise* to beam *Star Trek*'s Captain Kirk to some planet's surface and back to the *Enterprise*. Humankind may not be overtly aware of it, but it is getting itself ready, technologically, to do more than just project holographic images through spatiality; it is getting ready to transport material objects through it. There is a surreptitious logic underlying these imaginings, for humankind instinctively knows its biological imperative: that intelligent life must continue to evolve somewhere in the cosmos when earth is no longer a viable place to live.

But how do we reconcile a quandary such as the one we are contemplating, where one spacetime-transcending hologram, through a quantum-like biological interference, superimposes itself onto another unique spacetime and both simultaneously coexist? How does biology connect all this holographic travel—and seemingly more efficiently than light does? It is important to remember that one spacetime does not obliterate the other; it does not eliminate or threaten the invaded real-time or the conscious beings that inhabit it. Again, this goes against the Copenhagen grain of quantum theory, which stipulates that one of the two superimpositions must create reality by obliterating another's reality. Not all physicists subscribe to this version of quantum theory, however, and not all quantum "stuff" behaves in this manner. There are some quantum anomalies that exist, one of which we will now consider.

When Waves Become Particles

We are concerned with laser light and its beam-splitting effects and with biological and quantum dynamic paradigms; now we must find some way to unite them. To do so, let us take as an example the photon—-the elementary particle of indivisible quantum light that Einstein identified. Light can be described as wavelike and broken down into waves and particles. Originally, it was always imagined as being dual in nature until there was a way to harvest the elementary quantum photon—break it down to its simplest element. Eventually, scientists were able to produce many such elemental photons for use in different experiments. It was Einstein who proposed an experiment that had an individual photon choose between two slits. Then, guided by mirrors, it was forced to hit a particular area. After enough photons were fired at the slits, a pattern began to emerge. The pattern, which was supposed to show how the elemental photon chose one or the other slit, ended up displaying not one but two distinct choices that confirmed its original nature, wave, and particle. So, despite the fact that the photon had been reduced to one indivisible particle of light—a photon—it still chose to manifest itself as both particle and wave and passed through two slits instead of one. The original dual nature of the photon was retained, despite its artificial reduction into one element—truly a remarkable experiment at that time.

Today, physicists such as Eugene Polzik and his team at the University of Aarhus, Denmark, use a laser beam to entangle a cloud of atoms and change the state of an atom by transporting it to another place. The term that is used to describe this event is "teleportation," by which physicists actually mean that a change of place happens instantaneously with no intermediary intervention.[200] As mentioned above, physicists are now capable of teleporting small objects. The catch is that such quantum teleportation requires the receiving matter to be the same size as the object being teleported.[201] The fact is that physicists are on their way to teleporting matter that really would revolutionize spacetime travel, far beyond what I am suggesting here. So far, however, there has been no scientifically verifiable material evidence left behind by UFOs that they teleport matter, and it is a long way from teleporting a penny or data a few yards to teleporting flesh and blood. Therefore, for the time being, we are still bound to the holographic dimension, and so I return our attention to it.

Arthur Zajonc in *Catching the Light*[202]explains that quantum optics has revealed a phenomenological state for archetypal light that is truly revolutionary. It subsumes the premises of quantum theory itself—that an observer chooses what the outcome will be, and it can be only one. The problem is that physicists still have to make sense of this newly discovered anomaly of "wave particle-ness." Zajonc states, "[P]erhaps for light, at least, the most fundamental feature is not to be found in smallness, but rather in wholeness, its incorrigible capacity to be one and many, particle and wave, a single thing with the universe inside."[203] Keeping in mind these phenomenal quantum photon dynamics, I suggest that a holographic image traveling through spacetime might also be reduced to a photon-like indivisibility or wholeness. As a quantum of light encountering a fabric of spacetime around some alien planet, it would treat the crystalline substance as more than one slit. When it passed through two of these slits, it would produce two "interference fringes" of itself, as it does in experiments here on earth and as it does during the creation of holographic images by inventors when an object is split into two before it is reunited as a holographic image. Although it began as a quantum photon of light on earth, it would reproduce itself as both particle and wave once it hit an alien fabric of spacetime, so that it would manifest itself as both particle and wave, thus creating a translucent holographic image of whatever pictorial information had been sent.

Regardless of what kind of physical appearance aliens from other spacetimes might embody in their own worlds, because archetypal universal forms are contained within their biology, they would be able to respond to our environment by assuming the image of their choice, hence manifesting themselves in a form that, although perhaps remaining alien to some degree, would still be recognizable by us. This kind of scenario has been in use by science-fiction writers for some time now, and we do not lack species here on earth that, like chameleons, can radically change their appearances to match the environment.

Superpositioning Spacetimes

So far I have postulated that biological spacetime could answer many of our questions concerning where apparitions come from, whether from within our DNA or outside of it, in spacetime. What we need to do now is to further clarify how a quantum dimension exists within a biological notion of spacetime. The biochemical and quantum theory schools are replete with very different dynamics. Nevertheless, the two can be related because despite the organic nature of any chemical compound, it is always reducible to its quantum parts of wave/particle, whether organic or inorganic. At the moment, physicists are bent on proving that one all-encompassing formula exists that can describe everything there is to know about reality (the "theory of everything," or TOE). The problem is that their formula measures this "everything" mathematically, rather than organically and, therefore, in the end it comes to nothing more than mathematical conclusions. The organic, on the other hand, cannot be measured; it involves biochemical reactions that defy the laws of measurement; that operate under laws that belong to chemical and physical systems.

Under the laws of a biochemical system there is provided, at *every scale,* an easy image of crossover into another's biological spacetime. We have just to consider viruses, like the deadly bird flu, which enter into and hijack a cell's nucleus, forcing the cell to make viral copies instead of its normal ones. This can be visualized more easily when one incorporates the dynamics of quantum theory with a biological model, and that is why I have relied on this analogy. Further, taking Everett's notion of quantum theory and adapting it to our model allows for the possibilities of many spacetimes to coexist simultaneously and to interact with each other, despite the fact that they might be located billions of light-years away from each other. One has only to look at John Stewart Bell's mathematical proof that such nonlocal conditions exist and can communicate with each other over billions of miles, a phenomenon that cannot be explained away.[204] Einstein's view, in respect to the fabric of spacetime, was that if separate spacetime geometries existed, they would have to exist in a quantum-like universe.[205] More and more, despite the fact that Einstein did not approve of quantum mechanics, it appears that he is going to be proven right in this respect.

I have added an additional twist to this scenario by proposing that spacetimes, which are born out of biological substance, would not necessarily behave entirely like the quantum possibilities of the Copenhagen school, where real-time substance can only be created by the observer, and until observed, all possibilities remain "no thing." Biological spacetime, as part of a quantum mechanics model, would not proceed by these postulated, observer-created dynamics, by virtue of the fact that it is a biological model. In a biological model, spacetimes are compatible because they operate by the laws of biology, not physics. One spacetime observer would not obliterate another spacetime observer; rather, both would manifest themselves phenomenologically to each other as real. Suspended apparitions would occur that are not subject to a quantum indeterminacy.

If it is possible for other spacetimes to interfere with our own spacetime—and the phenomenal appearances reported by people throughout history suggest that it is—then we have to assume that our own spacetime also can interfere with another's spacetime. It is precisely this interference of our own spacetime by another spacetime, because of a transparent superpositioning upon our own, that creates the otherworldly apparitions reported throughout the ages. The question is, how and why does this happen? Since all planets, in our paradigm, that are suitable for developing life would be surrounded by a biological spacetime fabric of their own, they would always be participating in spacetime transgressions by virtue of their genetic biology and therefore engage in a natural affinity for crossover dynamics. In the same way that genes wander around the chromosome in biological nature, spacetimes would crossover and wander around in the cosmos. The same natural tendency to cross over in respect to electromagnetic waves is inherent in the micro- and macro-biological universe. The crossover, or the superpositioning of phenomenon, can also be thought of as unavoidable genetic transfers of information between spacetimes filled with life in the cosmos. Why? Because this is the way biological laws would work in a spacetime filled with electromagnetic frequencies.[206]

Our present fascination with a cosmos, built on a model of mathematics, geometry, and physics, has worked well for the evolution of our scientific imagination and still works well for it in a limited range of localized spacetime travel that can be measured without too much ad hoc guess work. Gathering information about spacetime, mathematically, is a model that has lasted for three millennia or more, beginning at least with the ancient Sumerians[207] and creatively addressed by the Mayas in Mesoamerica.[208] However, try as they might, physicists using this mathematical/physical model, which has given them license to theorize and then to posit black holes, cosmic strings, and antimatter as existing, to say nothing of the frolicking play of quarks and gluons, nonetheless still cannot figure out how we would remain the same selves after spacetime travel backward in time. There is the proverbial contradictory encounter with our grandparents, which, if it occurred, might obliterate our very existence. How could such backward— or for that matter, forward—travel in spacetime happen without our personal history, as we know it in present time, being changed?

Of course, it would no longer be necessary to adhere to, imaginatively, the idea of space travel into the past or future by a time machine, as our model works genetically, not mathematically, and it facilely rules out these past, present, and future phenomena by a new phenomenon, which I have dubbed "genetic time" or "an average of time." Past or future time is always within and/or without us in what we on earth refer to as real-time. This notion of an average of time would fit in nicely with Einstein's belief that there is no beginning or end to the cosmos, and it might appeal to those physicists who believe, somewhat like de Chardin, that in the end, all life in the cosmos dissolves into an entropic soup, as though life's mission is accomplished when it returns to some unknown unknown.

CHAPTER 7

REINCARNATION, INHERITED MEMORY, AND THE GENOME

In chapter 5 I argued for a biological universe that included crossover spacetimes. I made an analogy to Barbara McClintock's "jumping genes"; that is, to the transpositioning of genes and the superpositioning of spacetimes. I want to stay within these biological frameworks as we continue to examine the genetic nature of apparitions. This will include the already touched upon subjects above, while introducing some new but very familiar ones.

Reassessing Reincarnation

A good place to begin such a genetic examination of apparitions is to consider reincarnation. Most religions hold, as one of their basic tenets, a belief in life after death, which could be looked upon as a form of spiritual reincarnation of the person. Hindus believe that people are reincarnated as another human being somewhere on earth, depending on the merits or demerits they accumulated in their past lives. A reincarnation of this sort has nothing at all to do with genetic material; in fact, the biological is only considered when the soul must choose its new parents. One's inherited biological characteristics, therefore, do not count; only one's spiritual attributes accumulated in past lives matter. These good deeds ensure that you climb the ladder toward a better life on earth, with each reincarnation helping you to attain a purely spiritual existence, at which point you achieve Nirvana. And so it is that Indians of the Hindu persuasion can step over the sick and dying on their streets without remorse—after all, those people dying are solely responsible for their present condition, and they will soon have another incarnation, another chance to redeem themselves.

There are psychologists who are interested in assessing reincarnation, not necessarily to prove or disprove it but mainly because they see something interesting going on that they cannot quite explain. As they do with abductees, so too do psychiatrists resort to hypnotic techniques to regress willing people and elicit from them their memories of past reincarnation. In this way these psychologists provide us with a way to get at the root of what is going on when an abductee makes seemingly absurd claims. I happened upon a television series produced in Canada about a reincarnation researcher who films and/or tapes the proceedings during a regression session. She then later retraces the geographical places mentioned by the subject of where in the world her past lives were lived and who she was in that life. Then, at a later date, the researcher and her subject go to these recalled remote places in the world to see if the subject can remember having been there before. Some regressions that I saw were quite impressive, with people remembering odd foreign words and names of places that were no longer on the map (although the small villages still exist).

This kind of evidence, filmed and presented in situ, is difficult for a viewer to refute. The problem with the accessing of memories of past lives during hypnosis is the lack of biological connections between these lived lifetimes, which can take place all over the globe. This means that the soul is assigned to some genetically unrelated person in a very far-off region, implying that the soul, somewhat like Plato's soul in its ethereal dimension, floats around in what could be described as another spatial dimension, waiting for the right time and place to be born again. This is not unlike the ritual dynamic that takes place in the Tibetan Book of the Dead, where the soul of the dead person is gently persuaded to find itself another womb so that it can be reincarnated.[209]

If one believes in this version of reincarnation, then the entire genome project and the history of our migratory gene patterns would have no place at all in the scheme of who our ancestors were biologically. Scientific evidence would no longer be considered, and bodies would simply be vehicles for spiritual reincarnation. We have two different stories to choose between here: either we can adhere to the remembered ethereal history of a *soul*, as a subject tells it during a hypnotic session; or we can adhere to the biological history as revealed by a mapping of the person's genes. Of course, it might not have to be one or another; it could be a combination of the two. After all, we could have inherited more than a purely biological nature. We might have inherited a spiritual one also, unrelated to the bodily one. This gets a little complicated, however, and although I do not rule out finding some evidence for this dual inheritance, we will not proceed with such a dual notion.

In this chapter, then, my intention is to address what we presently know about how our biological inheritance works. Some of the material has been around for a while but is presently controversial again—like ancestrally inherited genetic traits for musical or mathematical genius—and would have us believe that all people are not born equal. Although there is much evidence other than genetic evidence to support these theories, my argument will be based strictly on genetic research, which I will use to anchor my theory that the genome is an inherited bundle of chromosomes, some of whose genes contain memories of past lives lived, including cosmic memories. My thesis, in this respect, is that we are indeed reincarnated beings but only to the degree that we begin our lives as a formal version of our ancestral genes, found in the mysterious molecule of DNA that is uniquely our own.

Jean M. Auel, in describing the brains of Neanderthal people in *The Clan of the Cave Bear*,[210] makes a statement that, I believe, contains more than a kernel of truth. She writes:

> And their memory made them extraordinary. In them, the unconscious knowledge of ancestral behavior called instinct had evolved. Stored in the back of their large brains were not just their own memories, but the memories of their forebears. They could recall knowledge learned by their ancestors and, under special circumstances they could go a step beyond. They could recall their racial memory, their own evolution. And when they reached back far enough, they could merge that memory that was identical for all and join their minds, telepathically.[211]

I am going to use Auel's insightful statement, written so many years ago, as a stepping stone to argue that in fact, such ancestral memory exists in all of us and is part of our DNA. In some of us, it is astutely developed because of how our molecules utilize this information to assemble themselves in our brains, either enhancing or dimming these inherited memories.

Cosmic Dust

But we are slightly ahead of ourselves in discussing our earthly ancestral memories. We need to first set the stage and to begin much more prebiotically with the possibility that life arrived on earth under the auspices of a meteorite, or in the form of a viral shard (viral shards simply copy themselves), or in cosmic dust—a view about the origins of life that until very recently was still very contentious, although scientists are now not adverse to its possibility. Christian de Duve, in his book *Vital Dust*,[212] to which I have already referred, contends that life began as cosmic dust and arrived on earth as "a cosmic imperative"—it just had to happen this way because the cosmos teems with molecular life. Now many years after de Duve's book was published, scientists are discovering the power of viruses—those somewhere between life and death entities—claiming that they may, indeed, be the forebears of the microscopic life that began on earth. In fact, scientists point out that most of the genetic material on this planet consists of viruses.[213]

NASA and its diverse body of scientists are pushing forward into outer space with probes that are equipped to send information back to earth. Scientists are also bringing back—successfully, in many instances—samples of material from the different planets or asteroids that have been targeted for assessment. We should have some definitive answers in the near future to confirm whether or not life exists somewhere else in the cosmos. Until then, we know that scientists dealing with rocks miles below the surface of the earth have found microscopic life. They believe that since photosynthesis is not involved, these microbes evolved because they incubated in and fed on gases emanating from the earth's mantle. If this extreme form of life is found on earth, then the reasoning is, "why not on other planets?"

It is an enormous leap from the first particles of life that landed on earth to our human existent nucleus of DNA. Still, it is this cosmic connection that we cannot ignore as we press forward with our examination of genetic memories and ask if they exist. And if so, did some of them originate somewhere in the cosmos other than the earth? Unfortunately, this cosmic leap cannot be made until life of some sort is found on another planet in our galaxy or outside of it and confirms the speculations of biochemists. For now, we must theorize about the genome humans have inherited and the possibilities that may be lurking within it, and ask if there are millions of years of inherited memories that are deeply embedded within the genome—even extremely archaic ones—that connect us not only to animals but to the very cosmos.

Many theories abound, including de Duve's account of how cosmic dust might have found our planet friendly to its biological ingredients, bound itself to whatever constituted our earth then, and slowly evolved into prebiotic life, which in time developed into biological life. Biologist Steven Rose puts it succinctly: "life is inevitably autopoetic, self-generating, self-developing, self-evolving."[214] He tells us that test-tube experiments have shown that life could have "begun with an auto-ribozyme (a ribozyme is a form of RNA) which could [have hauled] itself up by its own bootstraps" and began self-replicating, although he points out that experiments like these do not necessarily supply answers for the origins of life.[215]

Earth scientist Robert Hazen also provides us with an account of the different experiments to synthesize RNA and thus the possible beginnings of life.[216] He concludes his book by summarizing three different theories of how life might have begun on earth. I cite just his headings: a) "Life began with metabolism, and genetic molecules were incorporated later," or b) "Life began with self-replicating molecules and metabolism was incorporated later," or c) "Life began as a cooperative chemical phenomenon arising between metabolism and genetics."[217] Only de Duve overtly, albeit also theoretically, embraces life's cosmic roots and leaves us with deep feelings of cosmic connections. Because of his own confessed connections to a meaningful cosmos, it is worthwhile quoting him:

> If the universe is not meaningless, what is its meaning? For me this meaning is to be found in the structure of the universe, which happens to be such as to produce thought by way of life and mind. Thought, in turn, is a faculty whereby the universe can reflect upon itself, discover its own structure, and apprehend such immanent entities as truth,

beauty, goodness, and love. Such is the meaning of
the universe as I see it.[218]

With theories abounding at this moment as to how life arrived
on earth, there are some biologists who claim that this is no mystery at all; that the answers can be found within each one of us.
Joel Achenbach points out that chemical reactions are taking place
inside of us, all the time, allowing cells to convert energy and atoms
in molecules. Just eleven small carbon molecules could have provided the chemical reactions necessary for the development of bio-
molecules. Out of these biochemical reactions, amino acids, lipids,
sugars, and perhaps even a molecular gem such as RNA could have
evolved. As Achenbach writes, "a kind of natural selection applies
even to the world of geochemistry."[219] And it is this natural selection by the molecules and minerals on the earth that gradually produced a molecule from which all life on earth evolved.

Trans-dimensional Experiences

Graham Hancock, who is not a scientist, writes about trans-
dimensional experiences in his book *Supernatural: Meetings with
the Ancient Teachers of Mankind*[220] and takes another approach.
He links together shamans and science in a way that ends up supporting scientific theories that the mysterious "junk" DNA in our
genome might be a language in its own right, containing answers
as to why certain people are prone to otherworldly experiences.
Hancock's thesis is that DNA contains teachers *within it*, whose
manifestations during drug ingestion sessions have inspired
shamans throughout the ages, pointing to the sameness and perhaps a reason for the underlying similarity of ancient archetypal
drawings in caves. The messages from these so-called genetically inherited teachers are released by ingesting certain drugs,
although as Hancock points out, the biochemical nature of the
brain can be changed by other means too, such as meditation or

because one is naturally prone to such experiences due to genetic endowment.

The problem with Hancock's theory—and for that matter, the abductees' and religious visionaries' stories—is that these universal type of appearances dwell on story lines that go only around and around the same track, offering the same basic themes to the point of redundancy. Hancock has gone into great detail to convince us of their sameness, yet there is something disconcerting about this sameness of the appearances that we are discussing. We could say that because they are found in the DNA of all Homo sapiens' cells, it would be natural for people who share the same genetic makeup to report the same otherworldly experiences, if they were genetically so inclined. Then, there is the further consideration that an individual's DNA is locked up within his body, which itself is a circular system and not an open-ended one; thus, one is stuck with one's genetic inheritance. The supernatural messages received by these individuals do not appear to vary from one epochal age to another, the only novelty is that the stage setting represents a new milieu—the cultural environment of a particular age. We will come back to this circular theme of visions and consider whether it is of cosmic origin.

On the scientific side, Hancock cites, among others, Frances Crick's directed panspermia thesis, which Crick sets out in his book *Life Itself: Its Origin and Nature*.[221] Crick, who was one of the discoverers of the double-helix structure of DNA, was of the opinion, in his later years, that DNA was too complex a molecule to have developed by chance. It therefore must have been deposited on earth by some greater intelligence in the past, who seeded the earth with DNA using advanced technology. Crick speculates, for instance, that DNA might have arrived by spaceship as a bacteria sent by an alien civilization. Hancock also includes an interview with Rick Strassman, author of the book *DMT: The Spirit Molecule*. Strassman, who conducted U.S. government-sponsored research on volunteers, bases his book on this research.[222] The volunteers agreed to take DMT (Dimethyltryptamine) as part of an experiment and hence, even though sceptical, Strassman was able to study the nature of their hallucinations. He reports that subjects who ingested DMT, which contains some of the ingredients that ayahuasca does, consistently reported being in the presence of aliens in another very-real-to-them dimension or world. These volunteers convinced Strassman, a dedicated scientist, that what his data was uncovering was not based on hallucinations, as he originally believed, but on these volunteers' real experiences, which they had while in these other states. By and large, authors like Hancock and scientists studying the subject are on the right track; they are just slightly ahead of themselves (as are the other scientists cited above) when it comes to providing the hard molecular science to back their theories (although soft-molecular science, such as Hancock's, points to evidence from many other different directions and sources).

Ready-Made DNA

My own thesis differs from Crick's, who believes that DNA arrived "ready made" from space by a spacecraft that inseminated the earth, giving it a head start toward evolution. While this might be so, I am of the school that believes that life began on earth spontaneously, compliments of a heavenly body, such as a comet or meteorite hitting our planet and depositing bacteria and/or viruses that brought with them some of the six elements—carbon, hydrogen, nitrogen, oxygen, phosphorus, and sulphur (CHNOPS)—necessary for life to begin. As de Duve puts it, "These six elements, in a myriad of molecular combinations, make up the bulk of living matter. They were the main actors in the chemical birth of life as well."[223] New studies are confirming more than this cosmic dust theory, because scientists have discovered that meteorites that have fallen to earth contain carbon-rich (organic) molecules essential for life to begin on earth.[224]

In this sense, I am an evolutionist in so far as I believe that life evolved slowly and produced a vast array of species, among them many hominid species. At this point, however, I add another dimension to evolutionary theory. I find the abrupt transition from animal consciousness to our own minds and consciousness too great a transition to have spontaneously evolved in Homo sapiens in such a short span of time. My position is that when animal and hominid life had evolved and achieved a reasonable degree of sophistication, an interference with a choice hominid's DNA by an alien species occurred. Some, like Zecharia Sitchin, a Sumerian scholar (to whom we will return below), profess this happened approximately three hundred thousand years ago and involved using a part of the already existent humanoid genome to create Homo sapiens. I strongly suspect that something like this must be true (as Plato was fond of saying), mainly because of the creation stories we have inherited from the Sumerians, Jews, and Christians, and even the Mayas. Without this deliberate interference with evolving humanoid life on earth, it might have taken millions of years more to evolve the intelligence we possess today—or it might not have happened at all.

If the creation of present-day Homo sapiens was due to an artificially injected gene or tampered with by artificial insemination, it was done to introduce a compatible but more intelligent species' DNA to the DNA of the humanoid nucleus of some already-existing humanoid animal on earth. This would explain the appearance of fairly sophisticated, fluid cave drawings. If the same technique was implemented more than a few times, it might explain sophisticated civilizations like the Sumerians. Essentially, we know all animals are made out of the same molecular materials, with only a slight amount of combinations in their DNA codes, which are changed—although large chunks of uncoded DNA are common both to animals and humans for reasons we can only guess at the moment. Hancock's idea that junk DNA might represent an encoded language that we cannot yet decipher is a plausible one. Redundant as this junk DNA may appear, it might well be redundant because the encoded language contains the representational symbols of the unconscious, which carry memories of our origins, and only the assurance of repetition will safeguard these memories.

Since all Homo sapiens share the same DNA, it might explain why, when certain drugs are ingested, the same kinds of images appear to people throughout the ages, albeit clothed in different cultural contexts—images such as entopically produced geometric patterns, serpents, well-endowed pregnant women, theriantrops (half-human/half-animal creatures), and arrow-pierced stick people. (Hancock has provided us with ample illustrations of all these drawings in his book.) Still, as I have shown in earlier chapters, there is also another kind of appearance that takes place on an intellectual plane, quite different from the purely emotional plane or supernatural realm to which Hancock, for the most part, appears to refer.

My father's experience of a white lady was an emotional by-product of an uneducated teenage boy's mind. (He lived on the frontline, with WWI raging all about him. School was not an option.) When the timing was right for him, the image arose out of the earth spontaneously, probably from his archaic store of universal memories, not because he had read about the white lady or because it was part of his culture's mythology. Interestingly, the little men he constantly drew on the backs of envelopes and other scraps of paper look exactly like the images of fairy-people which Hancock provides us in his book, and they fit Hancock's descriptions.[225] There are many other accounts of white ladies that men have encountered, like that of philosopher Boethius, which relate quite a different, mature, intellectual encounter with the white lady. Some, like Swedenborg, had many conversations with angels and other-dimensional beings, which appear to have taken place in a hierarchy of ascending dimensional worlds and through intellectual channels (we should note that these visions and conversations began for him after a mental collapse).

Images: a Coded Language?

To dwell on the images produced primarily through drugs is to limit ourselves to just one way of assessing the images embedded in the language or code of the human genome. Admittedly, the capability of having such archaic supernatural-type experiences of images—of theriantrops and serpents and other bizarre creatures—continues to remain with us (as certain mind-altering drugs reveal) for reasons we have not yet been able to identify. Even the earliest civilization we know of, the Sumerians, no doubt inherited certain images that were of alien origin from their forebears' DNA.

During long hospitalizations, I experienced drug overdoses on two different occasions. Drugs were administered to me, mistakenly, minutes apart, by separate nurses. Each time, strange, ugly men flooded the room, the likes of which I later concluded I could never imagine my mind conjuring up, as they had never been part of my conscious or dream experiences. Often, when I was administered regular doses of drugs for pain, geometric patterns laced the walls of my hospital room. All of these visions appeared to me when my eyes were open. They were projected onto the walls of the room or hung suspended in space. When the bad experiences occurred, the reality (to me) of these ugly, bestial men was so terrifying that I beseeched my husband, who was with me on both occasions, to hold me and protect me from them. On another earlier occasion, while breast-feeding my newborn son in the middle of the night, I was again scared out of my wits, as an ugly, old, crippled man in chains appeared in the doorway and started toward me, as if to take my baby away. I recount these incidents to give witness to the fact that appearances do, indeed, live in my genomic memory and, I suggest, in each one of our genomes. They are capable of breaking through into a real-time situation whenever our bodies and brains are either exhausted, or drugged, or for some other unknown reason.

Did the theriantropic images manifest themselves to early humans because the mixture of genetic material was such a shock to the humanoid genome that these images are permanently embedded in our DNA? Is this what the existence of Neanderthals' similar but different existence and the small complement of their DNA in those peoples of European ancestry meant to reveal to us? Was their species' DNA also interfered with but didn't quite make the grade of Homo sapiens'? It could be that Neanderthals offer us a missing clue to the roots of our beginnings and that author Jean Auel is right about their having evolved telepathic instincts that connected them to their past ancestors and to each other, instincts that modern people have managed to suppress. While in this instinctive dimension—when we let ourselves go and are lost to the moment—-are we more in tune with animals' consciousness? After all, animals belonging to a particular species do everything telepathically between each other, using a variety of methods to communicate, including smell, sound, and song. Even plants and trees communicate with each other's roots underground or through the wind with their seeds, and they do better living in cluster communities than in isolated ones. In Homo sapiens' case, the instinctive memories that we inherited can never be erased, only suppressed. They are no doubt the reason that unconscious images, which are part of the collective inheritance, surface during dreams, trance states, and when induced by certain combinations of biochemicals. The fact that we can produce these images through administration of particular drugs offers scientific proof that interference with neuronal activity results in producing other realities.

We know that we are much like other animals on earth and that we differ from them very slightly; in most cases, the figure given is less than 2 percent. How animals "see" the world, we can only guess, although scientists are making some inroads; they have monitored the visual impulses received by a cat by wiring a computer directly to a cat's brain. Scientists' computerized information prints out the images seen by a cat's brain, which turns out to be the same as what we see.[226]

We no longer like to consider ourselves as animals, as we "have dominion over them" but our instinctive consciousness, at some level, is very animal-like. In fact, we are far more ruthless than animals, as shown in our dealings with them, with people, and by our plundering of the environment. Animals can be very devoted mothers and fathers, even though they live by survival instincts and eat their own young sometimes. True, they are territorial and fight for their kingdoms, but they kill mainly for food, not for ideas, as we do. Everything about human beings, who are so closely related to animals in DNA content, smacks of alien interference, which involved tampering with our humanoid earthly instincts. That, in the end—as each one of us still bears witness—did not quite work out on some basic level. Something is essentially wrong when it comes to being the kind, loving, and generous people that our spiritual teachers try to teach us we can be. Although they began preaching this to us a few thousand years ago and still preach it, it has little impact over the war mongers, then or now. Our instinctual nature to fight over territorial rights, as animals do, is embedded in our genetic material, and despite our lip-service attempts to appease this anomaly within ourselves, after thousands of years we still have not succeeded in overcoming it. Nor have we succeeded in taming the supposedly "intelligent" aspect of our humanity. We still fight over ideas of God.

That said, it is time for us to approach the genome and molecular memories in a way that enhances Hancock's and, like his, is still very much involved with DNA. More and more, science, when it is able to overcome theory, practically seems to be catching up with what is usually referred to as myth or science fiction. Yet is it? A case in point is that of Zecharia Sitchin, who years ago pointed out that Sumerians believed the earth was hit by an off-course planet, the collision causing it to become half its original size and creating many "moons" but capturing only one moon to orbit it, which became its largest satellite.[227] Sitchin is not a cosmologist, but he is able to read Sumerian tablets and so interpret for us the ancient Sumerian stories, particularly the epic of creation known as the *Enuma elish*. His account of how our planet was formed is not the later view of Mesopotamian theogony[228] but of cosmogony, which he contends the creation story in the *Enuma elish* describes. Today, science confirms similar cosmological dynamics. Astronomy's version, at present, is that the earth was hit by a massive body, thus creating the moon out of earthly material. Now, many years later, the Sumerian story of creation that Sitchin recounts has been vindicated to a great degree. Scientists, rarely interested in myth, remain largely ignorant of these or other ancient writings and do not throw accolades Sitchin's way. Still, the Sumerian story of the "planetary smash-up" does not end there. According to the *Enuma elish* this collision also brought the seed of life to earth.[229] Science is currently considering a similar scenario and asking if some stray comet or asteroid brought life to earth. It is perhaps the prime question about our biochemical origins for which science seeks an answer at the moment, and scientists are getting closer to the answer each day—for example, the recent announcement that one of Jupiter's moons (Enceladus) appears to have water on its surface.[230] And more recently, after crashing a probe into a moon's crater, NASA announced results that there was a significant amount of water found in their scientific samples.[231]

Sitchin also relates the Sumerian story of human creation, which in light of what medical science can accomplish today—implant a fertilized-in-test-tube human egg in a uterus and even correct errors in its genetic code before doing so—is not as preposterous as it sounded when he first wrote about it in 1978.[232] Later, in 1990, Sitchin revisited the material, armed with more recent scientific evidence to back up the Sumerian stories of how their scientists created an "Adamu," an earthling or "lulu" (lulu means man in Sumerian, although Sitchin points out that the nuance implied is "one of mixed blood."[233]).

The story recounts how Sumerian scientists (who were from another planet) created an Adamu—they mixed the material of the most intelligent animal they could find on earth with the essence of one of their males in a test tube and incubated the resulting embryo in one of their women, later known as birth-goddesses. In the beginning, as in our times, the results were far from perfect. As Sumerian accounts and drawings depict, at first they produced monsters, while at some point they even produced hybrid chimeras, with heads of animals and bodies of men. There was much trial–and-error, but apparently, Sumerian scientists persevered with their experiments and eventually came up with Adamus, who were created to be slaves to do the heavy labor of mining gold in southeast Africa some three hundred thousand years ago. Themes in this narrative are very familiar to us today—from artificial insemination to mining for gold in Africa. Similar themes exist in Peru and are represented on the steles that appear to depict gold-miner themes and in the other artifacts safely displayed in museums in Lima. In 2008 the oldest known gold necklace, dating to 4,000 BC, was found in Peru.[234] Again we are confronted with representational artifacts that point to a much too familiar repetitive cycle.

Nor do we have to look far afield to find other references to artificial insemination in the Old and New Testaments of the Bible. These stories relate how old women, like the childless Sarah (in the Old Testament) and teenage Virgin Mary (in the New Testament), were visited by angels who announced to them that they would bear a child (in Mary's case, God's child). What do these stories have to say to us? (We should note the loss of public interest in the virginity of Mary, which, when I was growing up, was considered to be either a mystery or a joke.) Should we take these stories seriously today, in the light of the medical technology we possess? Should we reassess their inherent values and reconsider the possibility that our own genome may have been artificially created? The stories are not as far-fetched as they once seemed to be. Neither are the depictions through the ages of chimeras, given the fact that human/

animal hybrids are being produced for medical research by manipulating stem cells across species' boundaries. These chimeric animals, however, have only a small amount of human genetic material
transplanted into them, mainly to experiment with growing replace-
ment organs, but who knows how far scientists could go with this?

Is it possible that we feel divine, special, and superior to other
animals on this planet because we have inherited, through the
superpositioning of genetic material (what some religious writers refer to as "divine sparks"), alien traits in our DNA that may
have refined our bodies as well as our brains, allowing us to speak,
think, and wonder; allowing us to develop rationally and take com-
mand of all the animals on earth? Someone as renowned as Francis
Crick, who believed that DNA was not of this earth—that it may
have arrived under the auspices of beings from another, more
evolved planet by means of a spaceship—doubtlessly did not read
any of Sitchin's work on this subject. Crick believed that the complex DNA molecule found on earth could not have evolved out of
the primary elements floating around in the cosmos; that the odds
against this happening were "1 followed by 260 zeros."[235] Christian
de Duve, on the other hand, believed that life was primed to happen
spontaneously from biochemicals floating in the cosmos and could
emerge in as little as two hundred million years, anytime it found
conditions suitable on some planet.[236]

Since we really do not know how or where our DNA originated—all we can do is speculate—we should perhaps pay close
attention to our cache of ancient stories and our brain's ability to
reproduce images for us, either during dreams, drugged states,
hypnosis, or spontaneously, while we are in a normal state for reasons we do not yet understand. What many of us seem to have concluded is that we are more similar than different from each other
and even from animals, and new genetic evidence is proving that
this is so.

With more and more being discovered about DNA, Robert
Sapolsky, professor of neurology and neurological sciences at

Stanford University, offers yet another perspective when he writes that we can no longer assume that the differences between all life on earth lie only in the genetic code; that errors in it are caused by deleting, inserting, or straight mutation. For example, with the decoding of the chimpanzee genome, no new information was reaped as to any differences in their brain's anatomy from ours. Accordingly, he concludes, there is no particular brain gene. This neuroendocrinologist reports that we have known for some time that the chimpanzee's brain is smaller than ours. The less than 2 percent difference between humans and chimps, thus far, appears to be not so much in the deletion, insertion, or mutation of the genes themselves; rather, it is in the relatively few genes that regulate the amount of division neurons undergo. The human brain has one hundred million more neurons than a sea slug's brain and considerably more neurons (presumably three times as much, since it is three times bigger) than a chimpanzee's, indicating that quantity not quality might produce the essence of humanity.[237] Sapolsky's is an interesting theory; still, we must ask how this radical human essence got into the quantity. After all, Neanderthals had even bigger brains than humankind and their "essence," whatever it may have been, remains a mystery.

Focusing primarily on the language of DNA may not reveal, after all, the answers to why we see appearances and visions in our dreams and enter into other seemingly real worlds. If these comparative findings between humans and chimpanzees prove to be correct, then perhaps we need to focus on the many more neuronal activities that are going on in the human brain and, in particular, on how they affect the visual brain in respect to appearances. These questions and other related ones are being asked by neuroscientists, like Semir Zeki, who studies the visual brain and how cells in what he refers to as visual models in the brain respond to the stimulus of light.[238] How neurons create the appearances, which humans have reported seeing throughout the ages, cannot be readily addressed, because scientists are reluctant to deal with ethereal images. However, because these images originate in the brain, they are born of substance, just as consciousness is, and therefore must arise from substance and cannot be ignored.

Bridging Gaps

Until we bridge the gap between substance and essence, we have to continue asking difficult questions concerning images appearing in dreams and images that are projected onto our own spatial environment. At the risk of being repetitive, we must ask again, are all of these images necessarily entopic? That is, do they originate in the brain, or do some of them belong to the cosmos itself? What role do emotion and feelings have in creating them? As Antonio Damasio has argued in two books, images are created through our feelings and our bodily sensations and only then become concepts.[239] Kant, who approached these questions philosophically in the eighteen century, states:

> To neither of these powers (intuition or concept) may a preference be given over the other. Without sensibility no object would be given to us, without understanding no object would be thought. *Thoughts without content are empty, intuitions without concepts are blind.*[240]

Damasio takes Kant's famous lines and puts biological clothes on them. I do not, however, plan on delving into neuronal or molecular activity in order to address the biological nature of appearances. This is a task better left to experts. Rather than enter into these molecular realms and their complex dynamics, we will stay within the macro boundaries of genes, wherein lie the possible traits involved in producing appearances.

Genetic frontiers are breached continually as researchers discover more and more about our genome. There is much speculation presently underway in molecular paleontology concerning the role of pseudogenes in our genomes, discovered recently in the genome's dead matter. Pseudogenes can be thought of as biological fossils littering our chromosomes; they are reflective of a genomic evolutionary time scale. Until recently, pseudogenes were dismissed as the remains of disabled genes long dead. Now, scientists have discovered that not all pseudogenes are inactive, as about 10 percent can be "resurrected"; that is, turned back on to create functional protein products. Exactly what role these resurrected pseudogenes play in the genome is still largely a mystery. What we do know is that they contain the history of our genetic development.[241] The kind of language used to describe pseudogenes is laced with metaphors, which strongly suggests that somewhere in our genomic history, remnants of personal ancestral memories no doubt abide. Could turned-on pseudogenes some day help explain why certain people see otherworldly images more easily than others?

Cosmos: a Biological Conduit

In focusing on appearances, I have stayed largely within the boundaries of human biology, although as stated in the previous chapter, I believe that the cosmos is also a biological "animal" and that images made anywhere in the cosmos can—and do—crossover into our own 3-D dimensional space. This dynamic crossover of a cosmic dimension into a human one reveals itself by taking advantage of certain atmospheric conditions, probably enhanced by the receptive biochemical conditions in a susceptible person's brain. Nonetheless, whether cosmic or earthly, these otherworldly dimensional realities remain in the realm of the biological. In Great Britain, a study commissioned by the government and completed in 2000 (whose author or authors remain anonymous) claims there are no such things as aliens or flying saucers. The conclusions are that most probably these sightings are attributable to weather or electrically charged conditions or to meteors—some effects are well known and others, not so well known. They might also be connected to plasma-related fields that can tamper with a person's temporal lobe and even the brain.[242] Actually, these findings bode well for my own theories as to why people see appearances in our own time and space; that is, why they see appearances caused by these cosmic crossovers.

Again I return to the theme that if we can make plasma screens of multicolored cellular light by using cells that contain colored gas, which when ionized release ultraviolet radiation to form a plasma that reflects these different colors when light shines through it, then surely the cosmos can create its own version of a plasma screen. The study conducted by the British government only guesses at possible atmospheric causes. It could be physical or electrical and/ or magnetic phenomena that create these appearances, and such guesses do not cover the wider question of why people relate to appearances as adamantly as they do. Because the findings in this document underline the unsubstantial nature of these appearances, the British authors do not believe that there is any danger of a collision with a solid object on earth, a finding with which I also concur. However, to throw more light on the subject of dimensionality and crossovers, I need at this point to introduce a more abstract dimension. If appearances do not have substance, a good way to examine whether or not they are simply illusions of consciousness is by turning to physics.

Theoretical physicist Lisa Randall firmly believes that hidden dimensions exist and sets out to show us why in her book *Warped Passages: Unravelling the Mysteries of the Universe's Hidden Dimensions*. Her dimensions, it should be known, are not biological.[243] Randall, employing the notion of Einstein's fabric of space-time, refers to these hidden dimensions as "branes," which stands in for the word membranes and hence is reminiscent of biological material. But this is not what she means. You will have a general idea of what she is attempting to portray by these hidden dimensions if you have ever had a CT scan and have seen the x-rays of how it produces slices of your body; better still, think of a sliced loaf of bread, she tells us, branes are separate to each other. They are unique dimensions that cannot communicate with each other. They are located on a huge cosmic membrane. Common to all branes is gravity, which permeates the entire cosmos. We live in a 3-D brane, and although we cannot envision or communicate with these other higher dimensions, they are nonetheless there, tightly packed into minute spaces within our own spacetime. Randall does not rule out the possibility that extra dimensions could extend infinitely without our seeing them.[244]

As one reads about tightly packed hidden dimensions, one cannot help but think of the tightly packed double-helix structures of DNA, containing forty-six chromosomes that host all our genes and traits. This is an interesting if somewhat loose analogy, but I cannot help juxtaposing the ideas of physicists' hidden spatial dimensions with my own biological notion of essential space that I have argued might be found in our DNA. I approached the subject in my previous book when I asked the question of whether or not feelings for spatiality—of being in a different spatiality that arise in us while floating in a 3-D work of art—do so because we are connected to an essence of space programmed in our very genomes.[245]

In that book I argued that the essence of space might be located on a gene in our DNA. I did not know then and still do not have any idea if or where it might be found in the genome. In geneticist Dean Hammer's book *The God Gene: How Faith Is Hardwired into Our Genes*,[246] he lays out the results of his research to identify the *genetic nature of spirituality* and offers the thesis that it is found on the VMAT2 gene.[247] His research amply addresses the many facets of why he concludes such a gene exists and why it is found on this particular gene, but it does not rule out the participation of other genes. I also pointed to Eugene d'Aquil's and Andrew Newberg's neurological research into what areas of the brain light up or remain dormant during meditation or spiritual exercises.[248]

Their research sidesteps genes and instead relies on pinpointing where the mystical nature of the brain is located. Newberg and d'Aquil point out that this overall striving toward a higher principle, whether it be Hindus' Brahma, Buddists' Nirvana, Christians' and Jews' God, Islam's Allah, and so forth dominates all cultures throughout the ages. In looking for a reason why this may be so, they came up with a very Darwinian answer—that this spiritual need to look to the heavens for answers as to why we are here on earth may be a survival strategy that was hardwired into our neural system.[249] Also of interest to us is d'Aquil's and Newberg's suggestion that if Jung was right about universal archetypes, then we must look to the archaic part of the brain and its primitive cognitive organization, which includes the brain stem, midbrain, and limbic system. It is this region that modulates "repetitive motor behavior and affect" and has to do with "the maintenance of physiological homeostasis. In short, such evidence would strongly support, at least in broad outline, Jung's archetypal hypothesis."[250] We will come back to the notion of neurologically wired-in survival and repetitive motor behavior below.

Not much is known about genes and memory, but more and more scholars are now studying how this connection might be made.[251] When scientists study memory and genes or molecules, they focus on the making of or loss of memory, long- and short-term. At this point, we are primarily interested in what is known about genes on chromosomes and about how our inherited genetic memory might fit in—and why appearances arise at all. I am betting that our DNA has bundled into it important information from the past and possibly the future of our cosmos and that the answers I seek here—why appearances appear to us in our four-dimensional space and time—will lie in the traits found on certain genes. The way in which traits are hardwired into certain circuits might explain why there are people who respond easily to archaic memories and can tune into cosmic appearances and other dimensions and why others cannot.

Genetic Memory

Genetic memory, as I am proposing it, is an inherited memory we all share. It is there from our cosmic beginnings because it comes from the cosmos itself. It is found in our genes and not necessarily only in a limited few. Each one of us is made from an identical package of DNA, differing only in traits. DNA is in the nucleus of every cell and contains all the information needed to create a unique individual. As already stated, I am not so much interested in memory research that aims to discover the protein molecules that can rescue the kind of memory loss one gets with Alzheimer's disease,[252] although without a doubt this kind of research helps and aptly confirms that we make new memories all the time and continually draw on the long-term memories we already have made. My immediate question, however, has to do with whether the memories we make will die with us or whether some of the new ones we make while alive are passed on to the next generations through our genes?

Daniel Dennett, a well-known author and atheist, argues in *Breaking the Spell: Religion as a Natural Phenomenon*[253] that "memes" (Richard Dawkins' word for propagating ideas), not genes, are responsible for keeping the memory of religious leaders in a community mysteriously alive throughout generations. Stephen Pinker, on the other hand, in *How the Mind Works*, argues that genes are adaptive or by-products of adaptation in what his colleague Steve Rose claims is a disconnected modular neo-Darwinian approach.[254] Rose, in *Lifelines: Biology beyond Determinism*, believes that we need to take a holistic approach to our genes and consider all the neurobiological dynamics going on, molecular and neuronal, as well as the social and cultural implications of our environments. We are more than the sum of our genes. Genes do not predetermine our social destinies; rather, we owe to them our creative intelligence, which is a by-product of our genomes.[255] All of these arguments offer interesting data and serve to illustrate that today, scholars in many disciplines are pursuing theses that include biological pathways or models. Even technological inventors delight in taking their models from nature; in fact, very few innovations these days do not borrow from it.

It is highly unlikely, although not impossible, as the science of epigenomics reminds us, that we pass along to the next generation the personal memories that we, as individuals, make. This said, there is more and more evidence that suggests that our genomes do not come to us tabula rasa but that our genes come imprinted with our ancestor's experiences and memories. Jung referred to the kind of genetic memory that he posited as the collective unconscious, although he did not examine this from any biological or cosmic perspective. Staying well within the psychological boundaries of his discipline, he examined cross-cultural symbolic values when describing archetypal dynamics in respect to the collective unconscious, particularly as found in dreams and in the mythologies of indigenous peoples. He believed in inherited psychic factors, stipulating that archetypes of the unconscious had no content, only form, at our birth. In other words, we came into the world wired for form, but until we experienced life, these forms remained empty—they had no content.[256] Jung's approach is similar to Kant's, Plato's, and Damasio's.

There is much controversy surrounding genetic inheritance, but we have come a long way since Jung identified the archetypes of the unconscious. Thus, to ignore the roots of what we know about our genetic inheritance (because of the threat of being labeled racist) is just plain foolish and detrimental to the scientific approach. Current scientific evidence points to the fact that the genetic inheritance of traits is possible and that people around the globe are not born with equal talents or desires. Some are born with traits on genes that have evolved in specialized ways over generations and that excel in certain areas, and there is plenty of statistical evidence to prove that nations of people who stayed together in limited areas and bred selectively within their own group have developed and fine-tuned specific traits.[257] What applies to groups of people and individuals applies to genes, too, which do their own choosing, activating either inherited female or male traits in the newly created genome. Most female animals, as almost every TV documentary on the subject reveals, do this kind of gene-selecting automatically by fending off all other males, until the male of their choice arrives. The human species is not any different in this respect (except in certain cultures, where females have no choice), whether this involves individuals choosing mates or nations closing their borders to foreigners, as is happening now. Humans are choosy people, but it would be folly to think we are all so different. In fact, quite the opposite is true.

Some scientists who study neutrons go to extremes by stating that we may all be descended from neutrinos. But we could not begin with these invisible, cosmic, ghostly particles, even if we wished to take our genetic argument that far back in evolution, as there is no concerted research in this area. Our genetic roots are closely bound with the cosmos, and the best guess at this point in time is that all life on earth developed out of cosmic particles and that these particles, which speak to us in images, contain cosmic information. Still, it could be that neutrinos can provide us with a clue as to what paradigm to assign to them when it comes to figuring out just what creates consciousness. Can consciousness be compared to invisible neutrinos that only in decay become matter? Do we become conscious when electromagnetism, or something like it, sparks this invisible dimension of consciousness into life, forcing it to decay into awareness? Are we just conscious flesh and blood "neutrinos," the leftovers of cosmic consciousness? The nature of consciousness and what it may be, however tempting, is beyond the range of this book, although there are reams written about this subject. Instead of pursuing the analogy of neutrinos to consciousness, we need to continue our discussion about the roots of our genetic memories, which must be addressed if we are to appreciate the origin of appearances.

The premise that I am offering—that genetic memories exist within our genome—has been well documented throughout the ages, both through the use of drugs (and the hallucinations they induce), by our existing religious factions, and most importantly, because of our feelings for space. Nobody would contest the fact that the human race, from time immemorial, has been continually fueled by the notion that we are spatial creatures with a special connection to the divine realm above, and we are less connected to the fact that we are creatures of the earth, although indigenous people try to bridge the gap by paying special reverence to a nature that is animistic. Because we have a pretty good idea of the biochemistry of our brains, and which drugs re-create which effects, and where all this happens in the brain, we need to determine on which chromosomes and genes the spatial dimension might be found—it is this essence of spatiality that might help us to understand more about "appearances."

Appearances, ostensibly, can be evoked through more than one venue; this is the reason why we have examined both the emotional and intellectual plane. We should, however, acknowledge the paranormal experiences that people have reported throughout the ages, as well as the people who seek to contact their dead relatives or friends, with or without a psychic medium's help. To me, this indicates the need to appease feelings for the person who has died. These feelings could be categorized as experiencing *psychic* phantom pain, not unlike the phantom pain experienced by people who continue to feel pain in their lost limb. These kinds of feelings of crossing over to other dimensions are a natural part of our genomic inheritance and have to do with individuals whose feelings for their loss cannot be erased from their brains. Such feelings depend more on the immediate biological makeup of the individual but can still provide us with clues as to why appearances of ghosts and haunted houses are such popular subjects, particularly in the entertainment media.

To this end, and beginning at what may appear to be an antithesis to anything biochemical, I want to further pursue in the next chapter physicist Randall's ideas of extra dimensions and her notion that we should relate to them as "warped passages." Can this mixture of purely theoretical dimensions, when added to the biological dimension, help us understand why either human-like or creature-like appearances, seemingly from other dimensions, have appeared to our forebears and continue to appear even today to people from all walks of life?

CHAPTER 8

ROLLED-UP DIMENSIONS

Space and Time: an Illusion?

I will begin this penultimate chapter with Lisa Randall's conclusions that cite references to other physicists' remarks, which state that "space and time may be doomed," or that "space and time are illusions," or that space and time may simply be emergent properties of some other as yet unarticulated theory.[258] She herself admits:

> Despite the impressive physics developments of the last few years, we don't yet know how to harness the force of gravity or teleport objects across space, and it's probably too soon to invest in property in extra dimensions. And because we don't know how to connect universes in which you could loop through time to the one in which we live, no one can create a time machine, and most likely no one will do so any time soon (or in the past).[259]

Can Randall's observations add further insights to our own biological ones? Working within a framework of theoretical or even
practical physics does not bring physicists closer to an elegant solution through which to express multidimensional or other worlds.[260]
Randall often reiterates that the problem lies with the Planck scale
length, which hinders mathematicians' ability to penetrate any further mysterious depths. While this occurs with physical models, it
does not necessarily apply to biological ones. I suggest that these
physicists' solutions to their problems with multidimensionality
lie hidden within the genome itself. Building a time machine will
probably never work in our cosmos—all the physical formulas
physicists have produced so far confirm this.[261]

Hence, are time and space illusions? It certainly appears that
they are. Even the cosmos itself, with its Great Wall of galaxies,
goes out of its way to remind us that however clever our calculations and formulas and explorations of space, we may really be
dealing with an illusionary time and space.[262] Only our genomic
material knows for sure, for within each of our DNA we may hold
the answer to the mystery of time and space, which we have inherited from our forebears, who live forever within our genes. In fact,
the moment of creating another human being who can think and
wonder and imagine is not unlike the "orgasmic" moment that is
described by scientists' big bang theory. This moment was recently
revised by physicists, who now claim that their new measurements
reveal that the big bang happened much faster than earlier theories
allowed.

The new theory has inflation occurring in a trillionth of a second
after the big bang, during which time all the seeds for stars, planets,
galaxies, and everything else in the universe were sowed.[263] On the
other hand, evolution might not have happened as sanguinely as
it is presented by Darwinians, nor might it have been a one-time
event. There is an alternative scenario; Stephen Jay Gould's contin-
gency theory comes to mind.[264] Our evolved human genome might
be a contingent event never to be replayed again. As might our big-

bang cosmos. Indeed, Gould's scenario might be the answer to the physicists' dilemma as to how the big bang came into being. Was the cosmos just evolving from an earlier state? Was it the birth of another newer one, perhaps more complex than its parent cosmos and based on some other unimaginable contingent event? If we think of our cosmos as a biological one, we could ask if something artificially inseminated an existing cosmos, causing our particular big bang to occur and creating our cosmos.

When physicists reach the limits of their theoretical multidimensional models, when the details they are working with are so tightly packed and minuscule that they cannot even be imagined in a thought experiment, then physicists are not dealing with something tangible but with *essences of dimensionality*, in the same way that one deals with an essence of spatiality. However, there is a difference. The essence of spatiality that we have touched upon in this book so far—spatiality or other worldly dimension of space in which visions appear—is not produced by physical formulas but is based on the foundation of almost identical stories told by a myriad of witnesses throughout many epochs of evolving consciousness. While I do not argue that appearances have a physical reality (they do not), physicists seek to prove their imaginary theories tangibly—something they wish to achieve with the Large Hadron Collider (LHC), which is finally allowing them to test their theories.[265]

Hidden Dimensions and Genes

Interestingly, physicists often use metaphors to explain their theoretical models, using language like the "fabric of spacetime," "a bowl of Jell-O," or "woven into spacetime," with the most popular being the word "membrane" that Einstein used and which we are often asked by physicists to imagine as a sheet of space. All these metaphors suggest either material or biological substance. Randall believes that rolled-up dimensions can be located at any point in 3-D space. She states: "You might liken the points in extra-dimensional space to the cells in your body, each of which carries your entire DNA sequence."[266] Even though she does not take this DNA analogy any farther, she hits the nail on the head, so to speak, with this throwaway remark. I believe that the hidden dimensionalities sought by physicists lie within our genes, which, if imaginatively combined with the essence of spatiality, can provide us with clues about the origins of appearances and visions.

In the book I have assumed that visions or appearances appear to people in a four-dimensional spacetime; that is, in an Einsteinian model that is most familiar to us presently. No one who is kidnapped by aliens ever talks about experiencing a warped dimension—a space so bizarre that Randall cannot describe it. Even if there existed other dimensions that are distorted, the human mind, at this point in time, does not appear ready to construct its visions within them, nor does the cosmos project them at those of us who serve as its receptors in anything but the spacetime we live in. This leads me to believe that however different the little people or graylike aliens might be, and even though they might transcend dimensional barriers as they slip between the lines of heaven and earth, we on earth have not been programmed to imagine these appearances in another dimensional environment, such as the fifth dimension. But who knows what the future has in store for us? We might, indeed, find material particles that prove the existence of the fifth dimension and *then be able to imagine visiting aliens at home in these dimensions.* For the time being, however, appearances generally take on the existent cultural accoutrements that the human being is already familiar with and limited to.

If we accept these appearances and phenomena as originating in our DNA and believe our genes contain memories passed down from ancestors who experienced them, then, for example, we might carry within us memories of these aliens from space. This would allow some of us, like the abductees, to recall and visualize aliens tampering with our humanoid genes. These memories may be the reason we experience both a connection to earth and to space.

Memories like this may be the by-product messages from the gods living in our genes. This means that the key experiences our progenitors had while they were alive on earth are forever embedded in our genes. These memories need not belong strictly to an exclusive genetic line (although some might), because as in tracing our roots we are finding genetic evidence that attests to common ancestors worldwide.[267] With most people on earth today connecting to an idea of God and with most believing that they will go to a heavenly dimension after death, genetic memories may be playing the lead role in sustaining their beliefs. These feelings for the sacred other, for the divine, for God, cannot easily be erased from our genome, although many atheists refuse to admit this.

Cosmic Clues

The cosmic connection, the whole, is greater than the parts of our cosmically inspired accumulated memories. Messages from ancient spacemen known to us as gods may still be stored in the electromagnetic fields that surround us and penetrate our planet in the same way that the earth's magnetic field flows in a circuit of great waves of energy into outer space. Herein lie the clues to unraveling the great mysteries that spacetime no doubt contains. And astronomers continue to ask the right questions: Why is the intergalactic gas found in the space between galaxies magnetized?[268] Why does so much feedback occur between galaxies in intergalactic gas and is cosmic illusionary feedback becoming "a unifying theme in astronomy, seemingly repeating itself on all scales"?[269]

Astronomers also are deducing that the cycles on earth are reflected in the cycles of the intergalactic medium; that the precipitation of exploding stars that produce a galaxy gives birth to new ones within it; that intergalactic winds exist as halos around a galaxy, just as the prominences of solar winds place a halo around the surface of the sun in our planetary system. They believe that the prominences jutting out from our sun are a result of magnetic

activity, and they ask, "Could it be that magnetic activity dominates our galaxy's atmosphere too? If so, the analogy between
galactic atmospheres and their stellar and planetary counterparts
may be even more apt than we think."[270] Given all the questions
astronomers ask concerning magnetic activity, it seems reasonable
to wonder whether the human brain is a part of the electromagnetic
cosmic feedback system and to wonder, too, whether this explains
why some people are particularly sensitive to seeing appearances
that appear to them against a background of real-time, especially
when they ingest certain drugs, or when hypnotized, or in dreams.
It seems reasonable to wonder why so much of the global population, even those who may not experience visions or appearances,
still believe in related theories, such as other worldly dimensions,
or heaven, or reincarnation.

In considering these scientific approaches, I have done so in
order to demonstrate that however diverse they may be, they still
contain areas that overlap each other. Physicist Randall tackles
dimensions from a theoretical point of view. Her dimensions are
made out of geometrical points that *if* connected would depict
what these extra dimensions of spacetime look like. The trouble
is, Randall explains, that these extra dimensions cannot be geometrically connected and drawn because they cannot be imagined,
except as rolled up in minuscule atom-like fashion in our own
four-dimensional spacetime. Still, they are conceived by theoretical physicists as real, just waiting to be discovered through
their material particles.[271] Randall hopes that the LHC succeeds in
producing Kaluza-Klein (KK) particles, which she defines as four-
dimensional particles with a higher dimensional origin. If such
KK fingerprints are found, then the existence of a higher dimension (or dimensions) will be confirmed concretely, simply by the
material evidence left behind.[272]

If the experiments scheduled to be run on the LHC [273] prove
that leftover material (particles) from a fifth dimension exist,
then Randall's theorizing will have gone a long way in proving,

concretely, that perhaps in some other galaxy or galaxies, a different dimensionality exists, with our own wrapped up in it. Therefore, even in such a warped dimensionality, if conditions were suitable for life, a different life form might evolve to suit what is a different dimensionality. Nonetheless, these as yet unimaginable spacetimes would remain part of our cosmic reality; they would not extend beyond our cosmos. As earthlings we are limited to living in a four-dimensional spacetime and to imagining all these extra dimensions and creatures in them as also existing in this four-dimensional spacetime. Yet physicists' theories aside, some of us still believe that fairies slide in and out of the earth, along with white ladies. And we have inherited a myriad of underworld stories from our forebears—an underworld that contains another kind of imagined four-dimensional world, instead of a warped fifth or sixth dimensional one, as physicists suggest.

In exploring the cosmos, astronomers are providing valuable clues with their always newer information, thanks to the sophistication of the instruments and computers to which they are now privy. They are beginning to see more clearly, where once they saw opaquely or not at all. They have arrived at a point where they can prove that the cosmic background is not nothing but something, with many faintly outlined spider-web filaments running throughout it and bombarded with intergalactic gases and cosmic winds. They have confirmed that a pervasive magnetism exists throughout the intergalactic cosmos and possibly throughout individual galaxies. These facts suggest that it pervades the entire universe.[274]

Electromagnetism and Magnetic Bacteria

My own thesis proposes that we are either receptors or projectors of cosmic or biological memories—or both at the same time. It doesn't really matter how a "tuned-in person" responds to supernatural images. One person may be more adept at tapping into the biological images *within* himself, while another may be more adept at tapping into the cosmically produced images *outside* herself. There are, for example, people who are particularly sensitive to electricity and install surge protectors in their homes for protection against surges of it. Many people suffer a range of illnesses, including sleep disorders, infertility, multiple sclerosis, fibromyalgia, headaches, and fatigue, which they claim are caused by electricity. Childhood leukemia has been linked to power-line magnetic fields.[275] But while some suffer from electromagnetic effects, others seem immune to them, while still others, albeit a minority, appear to connect through them to cosmic appearances.

It is not difficult to understand how these appearances or projections can arise before a person holographically. Again, I revert to my contention that the cosmos can project illusions and our brains can become receptors for them or, conversely, that our minds can project them into space, with all this happening holographically through our visual systems. Consider, for example, that in 1958, Pope Pius XII designated Saint Clare of Assisi as the patron saint of television on the grounds that when she was too ill to attend Mass, she miraculously was able to see and hear it on the wall of her room. Saint Clare was born in 1194 and died in 1253. She was canonized in 1255 by Pope Alexander IV. Today, we must ask, "Was she gifted with the ability to project that Mass from within herself?" We have already examined how we create holographic images artistically. I believe that if we can create holographic images artificially, then surely our visual systems can make them naturally and project them either onto the walls of our closed eyelids, or the walls of our minds, or onto the walls about us, or onto the landscape itself. This is what Saint Clare was able to do. Surely, if a human being can do so, so also must the cosmos be capable of creating holographic images. Most probably we borrow the cosmos' dynamics in order to achieve the holographic projections when we create them artificially.[276] It would appear that everything we on earth achieve technologically is a gift from our cosmic inheritance, confirming Plato's Ideas floating above somewhere in the heavens reflecting themselves as inferior copies on earth.

In arguing that appearances are caused by an individual's susceptibility to electromagnetism and that an individual can be either a receptor or creator of holographic images, I have no proof, aside from that of the Catholic Church's belief that Saint Clare could see and hear what I would term as holographic images and that this is actually achievable with biological substance. However, the hard evidence is that science is catching up with my hypothesis almost as quickly as I can speculate about it. As an example, consider the following: In reading about magnetic bacteria in a science journal, Kartik Madiraju, a sixteen-year-old Montreal high school student, decided to experiment with magnetic bacteria to see whether they would create electricity in a generator. He went on to demonstrate that these bacteria could, indeed, produce sustainable electric current. Madiraju is planning to continue a career in science and sees magnetic bacteria as potential for becoming an alternative energy source.[277] Magnetic bacteria provide us with a simple example but one going in the right direction. These bacteria help us put some flesh and blood on electromagnetism and link it to our bodies and brains. But this is only one aspect of bacteria that is used technologically. Being developed are biological computers, photon-based computers, and atom-based computers (the latter two based on quantum computing), and they are being developed more quickly than one can keep up with.

Our most ancient historical texts and, more generally, our cultural mythologies do imagine other worlds that present us with weird creatures but, as already noted, they do not depict them as being in anything but the spacetime we know—although our science fiction writers and artists, through the power of the computer, are perhaps on the verge of breaking through the imaginative barriers to which our notions of four-dimensional spacetime have bound us. But even the weird creatures used in computer games do not go beyond what are already familiar monsters. Maybe Randall's vision that we are stuck in a slice of four-dimensional spacetime is destined to limit the ideas of spatiality and creatures of science fiction writers and designers. It is good to recall, however, that before Einstein came along to change our imagination to a relative notion of spacetime, we were entirely caught up in a Newtonian version of space and time, where time was constant and unchanging. Perhaps today we are ready for a different notion of space and time, but will it be a physicist's fifth dimension of space and time?[278] Even if we prove that the fifth dimension's particles exist, how would we depict this spacetime, let alone try to enter into a rolled-up, so far impenetrable boundary that lies somewhere within our slice of space? The best we might accomplish, in my view, is to enter into *an essential spatiality and time.*

Randall's idea of being trapped in a slice of four-dimensional spacetime can be used to explain why, throughout the ages, people see appearances of white ladies, angels, or other messengers from God, and why abductees visit other worlds by riding up beams or traveling down into an underworld, and why cave drawings of chimeras and existent animals are so similar throughout time. Are we simply stuck in this dimensional plane? If so it appears that we are destined to *visually* experience this phenomenal repetition ad nauseam, as thus far we have not been able to transcend this model. An excellent example of this was given in an earlier chapter when we cited Joseph Smith's vision of the angel Moroni, who appeared to him in exactly the same spot, dressed the same way, and relaying the same message throughout one particular night. Most religious visionaries see the same appearances throughout the course of their visionary experience, always dressed the same way, standing in the same grottos, and conveying the same messages.

Trapped in Cosmic Feedback

If we are truly trapped in a dimensional circularity going around and around like a stuck record, how does the cosmos figure into all of this? What role does it play? Is the cosmos a lame duck, cooperating with earthlings by playing back to us our collective genetic memories that have been projected into space like radio waves over thousands of years? Why would the cosmos engage us by projecting the same feedback of illusionary imagery, no matter when in time it originated? Or is the cosmos providing us with a way to break the code and finally realize that all these appearances and gods speaking to us or commanding us to worship them, feed them, sacrifice for them, and kill for them and to be very docile, moral, prayerful people is just the product of our genes, hence inviting our brains to recognize these images as such? Or has the cosmos furnished us with this redundancy of appearances in order to get some other message across? There are many questions to be asked concerning our relationship with the cosmos, and all possess a possible kernel of truth, if only we could find answers for them.

Whereas it is usually assumed that the cosmos was created during the big bang and the earth eventually evolved into a habitable planet within it, the question of what came first, the chicken or the egg, may not be as transparent as it seems. In respect to the cosmos, it could be that we humans are the chickens and the cosmic egg appeared only after we gave birth to it. Could it be that we earthlings are willing the cosmos into existence? An example of this latter paradigm is found in speculation about the possibilities of the existence of other planets in the cosmos. Consider Frank Drake, who co-directs SETI (Search for Extraterrestrial Intelligence Institute), who developed an equation in 1960 that pushed astronomers' knowledge about the cosmos to its limits at the time. His formula was purely speculative. Still, over the years, his every guess (except one) proved to be viable in his speculative equation and only the very last value has not yet been verified.[279] Granted, we are just talking about guesses and a mathematical formula, which suggests the values to use in calculating whether intelligent species live elsewhere in the cosmos on planets similar to our own. The fact that years later, Drake's formula has almost been verified makes one wonder whether all human speculation, when logically derived, leads to the answer that an individual is seeking.

More than our collective cosmic memories are involved here; there is an infectious spreading of "memes" or ideas—a plasticity at work in the minds of the human species that gives content to the cosmos and its mysteries, which more and more seem linked to the same paradigms and dynamics that human beings experience on earth. Animals live this plasticity through their instinctive cooperation with the electromagnetic fields of the cosmos, which, considering the diverse and special talents of the millions of species on earth, are awesome to contemplate. Humans live their plasticity with the cosmos through their intelligent questioning, cradled in instinctive speculation, a by-product of their animal inheritance—a consciousness that in the beginning must have been a lot like Jean Auel suggests.

So what is the possibility that we can transcend this holographic universe, as Talbot refers to it, and the circularity of similar images that make their phenomenal appearances to receptive people from within or without themselves? Are we creatures forever bound to this slice of spacetime, as Randall suggests? I have argued that there is yet another perspective from which to approach spacetime—that our genome contains a gene (or genes) for space and time in an essential biological way. This allows us to enter into and interact with other spatial dimensions but only through an essence of spatiality—in an ethereal reality that one can never touch but can proprioceptively feel.

So we must ask yet another question: Can we break through this double-bind situation in which we appear to be trapped? Although we are not exactly in a cosmic prison, as our illusions ostensibly inspire us to achieve new and imaginative heights, are we building up irreversible entropic energy through these repetitive narrative cycles? Are we creating a condition that will simply dissolve us into the "nothingness" that philosophers like Hegel write about, or will the explosive nature of this irreversible entropic energy propel our remaining particles into some new, as-yet-unimaginable dimensional formation at the expense of life? Some people fret that the experiments to find the Higgs particle might do just that—explode the world—but others are more optimistic and hope to discover a Higgs field, one that underlies all other particles and acts the way a Jedi knight in *Star Wars* would—as a carrier of the force.[280]

The Dynamics of Entropy

In the *Cosmic Circle*, Robert Langs, a psychotherapist, and Anthony Badalamenti, a mathematician, set out to prove that mind, matter, and energy are not only intimately connected but that they are one and the same thing.[281] They do so by charting the results of the conversational dynamics between different therapists and different clients. That is, they chart the *dynamics in immaterial systems.* Langs and Badalamenti claim these dynamics are just as subject to the *Second Law of Thermodynamics* as material systems are.[282] They use mathematical models and psychoanalytic methods to analyze and chart their results from five different dimensional perspectives, which they identify as "narrative imagery," "new themes," "positive tones," "negative tones," and "continuity."[283]

The measurements of these dynamics between therapists and clients include the role of time, the encoded information in the narrative, the amount of energy expended during the dialogue, the rotation from narrative to nonnarrative forms, the state of tension during the rotation between nonnarrative and narrative forms, and the energy retained or irrevocably lost in the created cycle. Their use of different formulaic approaches is too complex to begin to describe here; suffice it to say that by applying mathematical formulas to the dialogues between therapists and clients, Langs and Badalamenti arrive at statistical conclusions that do not differ, in essence, from the lawful natural dynamics applied to material systems. They demonstrate scientifically that there is such an immaterial thing as *psychic energy,* which is produced during a closed system dialogue between therapist and client. No matter which therapists and clients are teamed up in a session, the energy that flows between them adheres to *the mathematical laws of nature* that the researchers characterize as being in "the mental as well as the physical domain."[284]

Of interest to us is Langs' and Badalamenti's use of the dynamics of entropy and how it works in a "cosmic circle" toward the integration of mind, matter, and energy. When we add their notion to our paradigm of appearances, as being either self-generated or cosmically generated, it adds an interesting perspective. Being subject to the laws of entropy, such an entropically organized cosmic circle, allows us to interpret appearances as part of a cosmic psychic energy that permeates all of matter and mind. Langs' and Badalamenti's idea, that a repetitive cycle can keep entropic energy in equilibrium or generate enough pent-up irreversible energy to force a client to break out of the repetitive cycle onto new psychological grounds, is very interesting. If we do the obvious and make an analogy to a dialogue between earthlings and cosmos, the analogy suggests two things: either we are caught in a cosmic circle of perfect equilibrium and cannot break out of this cycle of appearances (this would coincide with Randall's view), or we are not. Might it be possible to break out of this present cosmic spacetime cycle through an accumulation of irreversible energy into some unfathomable and indescribable dimension beyond space and time that exists outside of what could be just be an illusionary cosmos?

Which will it be? Are we eternally confined to the circularity of the space and time that we experience now, or will we, at some irreversible point, be released from it?

If our genome is, indeed, programmed to stay within the bounds of this holographic cosmos that human intelligence and imagination might be creating, there are still many alternative venues to explore, such as the different facets of the visible and invisible (for example, the dark matter that we cannot see). I have no doubt that in respect to our cosmos, whether illusionary or not, we can expect to mentally conquer it—not necessarily by physically going to planets outside of our galaxy but by using mental faculties we have not yet explored. We may someday even imagine a way to physically enter a weirdly warped dimension and do so, providing physicists first proof that these dimensions exist. But for what reason and to what avail? Our destiny on earth, at this moment, seems to be written in the sands of time's circularity found in our DNA, something the Mayas knew instinctively about time when they envisioned the universe as the provenance of the time gods that appeared and disappeared at regular intervals.[285]

For the time being, it is wise to acknowledge and appreciate the value of appearances for the equilibrium that they render to the imagination and intellect. They provide the inspirational intercourse with the cosmos that is necessary in order for us to reach new heights, create new knowledge, and keep our cosmic circle flowing. Life in this four-dimensional cosmos is all we know, and better what we know than what we do not know, whether it is real, illusionary, or holographic. It is sad to think that we earthlings appear to be on track to systematically destroy ourselves and all life on the planet, leaving our planet as dead as the other planets in our solar system. No doubt, before we destroy our earthly home, environmentally, we will probably do what aliens did on earth— implant the seeds of our DNA on some other planet in our reachable cosmos and thus continue the repetitive cycle of life in a "this cosmic dimension," replete with appearances and all.

CHAPTER 9

APPEARANCES AND THE TECHNOLOGICAL INSTINCT

Having argued that appearances are purely biological in nature and are related to our genetic inheritance, which itself is part of the greater biology of the cosmos, in this final chapter we will consider what possible purpose appearances may have and if they are influenced by both an emotional principle and an intelligent one. Many people experience appearances that seem to appease psychological/biological needs, while others have appearances with intellectual content. Because we have examined appearances from both these perspectives, let us revisit the UFO phenomenon in order to shed light on what is ostensibly an almost obvious answer to why appearances still appear to people and a most important proof as to why we have this desire to conquer space—this strong sense of cosmic connections.

By linking appearances to the intellectual and scientific nature of our instinctual drive to conquer the cosmos, I must step back, more or less, into our present spacetime paradigm with this caveat: that this instinctual drive will never be on the right course until the speed of light can be overcome. All the same, the desire to conquer space, which I believe is propelling us to explore space, certainly provides us with a reason as to why appearances are, indeed, a legitimate part of our natural instinctual drive. This said, I will begin this concluding chapter by turning to philosophers, a psychiatrist, and a physicist, in order to set up our climatic mise-en-scène.

Philosophers and Appearances

As already noted, philosophers have long been preoccupied with appearances. As philosophers, they have approached the subject from an intellectual viewpoint of phenomena, of objects, and of things and laced them in abstract conceptual language, which for Martin Heidegger, a twentieth-century philosopher, simply became a "thing-in-itself." Such a thing, and its alluded-to formlessness which has no image at all, is difficult to imagine, let alone conceive. But this is not where we will begin; rather, we will get right to the heart of the issue by tackling what Heidegger referred to as a technological essence and I, as a technological instinct. Technology, I believe, is the most important by-product of appearances, although some would argue that the notion of God is most important—important because notions of God reinforce the appearances of UFOs. Again, we cannot deny that the feelings of wonder and awe for something greater than ourselves that emanates from the heavenly cosmos is evident in almost all philosophical discourse and is, of course, the basis of all our religions. I do not plan to analyze why we relate to a divine principle or to ask whether our DNA contains the answer to what God is—this is subject matter for another book.

Heidegger, in his earlier work (something I discussed more thoroughly in another book[286] and therefore will do so here in a very cursory way), saw technology as a destructive weapon in respect to the earth's systems. In his later work, technology harbored the "essence" of technology, which can be a positive as well as negative aspect within technology. Heidegger begins his argument by pointing out that technology can set itself upon earthlings, by acting upon them in a way that forces them to be nothing short of slaves. However, as he points out:

> [T]he frenziedness of technology may entrench itself everywhere to such an extent that someday, throughout everything technological, the essence of technology may come to presence in the coming-to-pass of truth.[287]

We are talking about technology per se, but Heidegger's intent is to introduce us to an essence of technology by which he means something quite different. There is an anomaly here, as the technological essence inspiring technology can enframe our worldview by controling and limiting us, or it can liberate us. Heidegger hedges his bets, not only on the limiting aspect of technological essence but also on the possibility of its providing a future breakthrough into an entirely new existence—a new dawn; indeed, "a new mythic structure of great poetic power." This new mythic structure, which more appropriately should be referred to as an "old mythic structure," is what I believe appearances represent. It is a mythic structure that originated not on earth but somewhere in the cosmos.

Computer Logic and Pure Technology

To understand what Heidegger means by a technological enframement that can enslave us or free us, we need not look very far for an example. Hugo de Garis is a super-computer circuitry designer who believes that computer systems will soon be running themselves, with human beings no longer in control. De Garis is busy creating circuits that will design other circuits, ad infinitum. Using a set of rules that already allows him to guide the growth of *simulated* neurons and synapses on a two-dimensional circuit plane, he believes the real breakthrough will come when three-dimensions are introduced to the circuitry. His key idea is simulation, and he is presently working on simulating a kitten, but his ultimate hope is to simulate as much of the brain as he can. Sceptics doubt that computer simulation can ever provide an environment rich enough to nurture the superbrain that de Garis wants to create. But de Garis remains nonchalantly adamant that simulation is the way to go; anything less than this approach would slow things down too much. He, like others, envisions ideological debates about whether these kinds of super-brains ought to be built but retorts with the quip that people will have to get used to the idea of ranking second on the intelligence scale, although he also admits: "I'm worried, too, because I don't want to be swatted like a fly." [288]

How do the super-computers being manufactured at this time work? What is the underlying logic they use? The article from which I reaped this information cites five examples: fuzzy logic; the expert system (deduction logic); data mining; the genetic algorithm; and the neural network (induction logic). [289]

These examples of the logic that will underlie a super-computer's capabilities are the reason why people like de Garis place their bets on the power of the computer and why they believe that human beings will simply have to adapt to the idea that they are second on the intelligence scale. The computer's ability to electronically shift through masses of information and to induce and deduce possible answers to some problem from its analyzed information far surpasses our own ability . We are told that computers are coming up with some interesting innovative possibilities that would take human beings, collectively, years of plodding through and experimenting with millions of possibilities to arrive at scenarios for innovative models that might work. Computers, it would appear, are now acting as prime movers of the imagination in a way that might never have been foreseen by human beings in the past. There seems to be two reasons for the computer's success: first, it can endlessly and tirelessly go through trial-and-error experiments; and second (referring specifically to the hybrid system called Engeneous), it can "discover reality from the bottom up," using a model-free method that opens up entirely new windows on understanding, per se.[290]

The implications are staggering, especially in respect to the computer's ability to take the imagination into uncharted areas of possibilities—but would they result in "a mythic structure of great poetic power"? And what does it mean to say that a computer that works model-free (unbiased) can come up with a new way of understanding the meaning underlying our existence? Is such a computer-based understanding possible? Can the computer take the human imagination into new dimensions of reality that will help us create a new reality? Will computers, being lauded as advancing the best possible model-free understanding of a problem and as offering fresh and innovative paradigms to our understanding, overtake the imagination of our inventors or technologists? Can the unbiased understanding of the computer provide the human imagination with more than has already been provided for it naturally in the past, when computers were not part of our world?

Interestingly, Heidegger is also enamoured with the ancient Greeks (Parmenides is one of his favorites), but Heidegger's vision of the gods, the fact that they have "flown away," and his use of all this mystical or mysterious language in relationship to appearances is somehow inadequate, perhaps because it fails to see beyond the Greek experience. Nonetheless, when Heidegger speaks of a breakthrough of the holy or the sacred from one dimensionality into another, he gives us a philosophical version—at least in my view—of what religious visionaries claim from their perspectives to be spiritual manifestations that break through into our dimensionality for an unknown spiritual reason. So we must ask, have the gods actually flown away, or are they really all around us in the appearances and visions that people continue to have?

This genre of breakthrough appearance is not without its counterpart in the Bible, where angels and the messengers of God must
also break through from some beyond to descend from heaven into
our earthly dimension. These heavenly appearances have had control over the spiritual imagination of humankind since our recorded
history. So while some philosophical experts may argue that Heidegger has left the Judeo-Christian model behind for a Greco-Ger-
man pagan one, it is also arguable that Heidegger is in the process
of recovering the dynamics of appearance experienced by Judeo-
Christian prophets and their encounters with angels, and messengers,
and gods—or at least he has set up the framework for this recovery to happen. If one reads Heidegger's dynamics from the point of
view of appearances being UFOs and of their breaking through from
their world into our earthly dimensions, it strikes one as uncanny
that his strictly philosophical approach fits in so neatly with the
reported dynamics involved in the UFO and alien phenomena that
have haunted the earth since ancient times. Not only do UFOs and
other appearances continue to taunt the earth's intellectual elite, but
they also have forced the earth's controllers—politicians, academicians, and scientists, who claim it is all superstitious nonsense—into
a close-minded framework in respect to the world of appearances.

Mysteriously postured in a Greek pantheon of ideas, Heidegger's idea of gods withdrawing from earth is also a core idea
in cultural histories. For example, the Incas were decimated by the
Spanish conquistadores, which the Incas mistook as the return of
their gods. Mayas continually waited for the return of the next recy-
cled time god in their carefully calculated calendar cycles.[291] Jews
still await their Messiah, and Christians await the return of Christ
to earth. Using Heidegger's language, it is not God but *aletheia,* the
truth hidden in the appearance of thingness, that we have to confront and identify. Heidegger believed: "Modern physics is the herald of enframing, a herald whose origin is still unknown."[292] In this
respect, postmodern physics might prove him right. He even hints
that the truth will go beyond the revealed technological essence,

but only when openness to the technological essence is accepted by those on earth.

Thus, to answer the question of just what technological essence or instinct might be, one has to ask what possible reason underlies the love affair humankind is having and always has had with machines? The notion we should underline here, as another important metaphysical aspect of our computer age, is virtual reality and the power it has in the shaping of our future. It is what will allow human beings to stretch their imaginations to encompass the domain of gods and appearances. By channeling these instincts for technology, we are now at the point where we hope to settle, in the future, on some other planet, but what for? Well, unconsciously perhaps, because we want to save our earthly race from extinction. There really is only one reason that I can come up with for this historical preoccupation with technology: we have taken this direction because the human being is born with the insatiable instinctual curiosity for technology inherited from his cosmic ancestors.

A Practical First Step with Virtual Reality

And so we fall back upon the mesmerizing image of a Platonic moving and evolving cosmos. Plato's cosmos and the ancient concept of eternal return is one that can be imagined as a constant returning to a beginning that is reminiscent of past beginnings. It is as though he wants to perpetuate an image of a biological cosmos that provides an intelligent, somatic form to seed itself, as the ancients before him suggested. As we begin to discern the different patterns and laws operative in the cosmos, we also can more readily come to terms with the reality of the symbolic laws proposed to us by nonlinear thinking. Physicist Andrei Linde propounds a self-reproducing inflationary universe.[293] He favors a fractal universe instead of a limited vision of an expanding fireball, big-bang universe. His thesis would have universes gestating and birthing other universes in a never-ending cosmic procession. But he proffers this caveat: a fractal universe is too difficult for the average person to imagine, let alone begin to understand. Because I agree with him about the difficulty of discerning such a fractal universe, I will not attempt to recapitulate his description of how it would work. I do, however, envision it as being yet another version of a moving image of eternity.

I call Linde's work to our attention not just to posit the plausibility of such a moving image of a cosmos but to demonstrate that many scientists today, by using scientific models, end up explaining a moving, animistic Platonic image of the cosmos. What is interesting about these modern-day paradigms is that they allow us to contemplate, in more scientific terms, the human/machine symbiotic instinct. It does, indeed, appear to be cosmically programmed to arise on a particularly lush, user-friendly planet and to evolve and naturally devolve and eventually to die, as cyclically as all life on earth does. Caught up in this cyclic cosmic paradigm of birthing and dying, it would appear to be our destiny as earthlings to pass life along, a destiny that has enabled us to arrive at a test-tube technology that will allow us to create new hybrid life of our own design some day. These, like us, will be hybrid beings, which are destined to enter into strange relationships with the cosmos and machines for reasons they too will not at first understand.

Our myths are indeed revealing to us that our instincts for the cosmos go hand and glove with our instincts for technology, and if we thoughtfully reflect upon these dynamics, we can accomplish a cosmic breakthrough that would not have happened without appearances, spiritual symbols, UFOs, aliens, other dimensional experiences, and so forth. This cosmic drive or energy in which we are immersed contains the mythical laws that create the technological story in whatever style or design necessary to stimulate the imagination in order to make sure that the cosmic project is kept alive and well-functioning. However mythological and illusionary scientific thinkers claim the UFO symbols are, scientists ironically continue to create the necessary technology to accommodate and give life to them. Scientists are the ones producing the spacecraft that will help pass along intelligent life in the cosmos—*they are the new myth-makers.*

Presently, we are focusing on another step forward in our cosmic adventure by simulating virtual reality scenarios that depict such spatial adventures. Since it is part of the cosmic schematic for human beings to submit to the cosmic drive, it is no doubt also natural for us at this point in our technological know-how to be preoccupied with virtual reality as it prepares the way for our further initiation into discovering the nearby cosmos.

Genes That Remember

Before I continue, I need to comment further on this underlying present-day erotic connection to the cosmos, one that I see as an important step forward because it potentially throws light on the cosmic/human connection. I suspect that people are unconsciously remembering, tapping into this genetically programmed desire to gestate and nurture cosmic life forms. In so doing, people are thus embracing a new-but-old idea of the symbiosis of the human/divine hybrid notion of being. They are opening their minds and hence inviting others to interact, not with some unknown aspect of the cosmos out there but with the *laws of cosmic energy itself.* Among the people who are being cosmically conditioned to accept responsibility for a new human/divine (hybrid) reality are the abductees who claim to encounter aliens and to travel intra- or extra-dimensionally to the UFOs, where this hybrid notion is reinforced. This, I believe, is the practical first step required to imagine interacting with an unknown environment that could help humankind leave our planet behind someday, when its resources have been entirely depleted; when it has completed its cycle and finally must be left behind as a barren wasteland in the same way that, perhaps, Mars was abandoned long ago. It also points to the necessity of creating hybrid life with the most suitable species on some alien planet to ensure survival of intelligence, per se.

Can it be, then, that the ancients' metaphorical *"deus ex machina"* (a machination that they constructed as a prop and had dangling in space in some of their theater productions) continues to inspire us today? Does it inspire us because we are the hybrid beings made by those who arrived on earth eons ago and set up shop to mine it for its gold, creating slaves (us) for this purpose? Were they the "techno-beings" we are becoming?[294] Since the mythologies of the ancient Sumerian, Babylonian, Egyptian, and Mayan realms bespeak a preoccupation with gods arriving by machines, it is reasonable to assume that our present-day preoccupation with virtual reality, which creates scenarios where people can experience simulated spatial conditions, is a natural and practical first step toward conditioning the mind to accept its role in the colonization of another planet. In essence, Michio Kaku's book, *Hyperspace*, is about amalgamating the mathematical formulas required for describing, at one shot, the basic forces of the cosmos—gravity, electromagnetism, and the strong and weak nuclear forces—as vibrations at work in a higher dimensional space. Kaku, a well-known physicist, emphasizes that physicists do not speculate idly on matters of space travel, as our technology is exponentially doubling every ten to twenty years, and we are on the frontier of breaking through into new horizons of harnessing the power of our planet and then the power of the sun—and then that of our entire galaxy. Who knows how far we can go? Once intelligent life reaches a certain level of physical control in the cosmos, it may become virtually indestructible—immortal.

In our remembrance of the cosmic instinct and in our reinforcement of it technologically today, can we find in this cosmic scenario the continuing purpose of the cosmic life force? As I have said above, I believe such a cosmic life force seeds intelligent life on certain user-friendly planets for the reason of passing along life in the cosmos. When a particular evolutionary point is reached on this or another planet (and I do not think we are the only intelligent creatures involved in this cosmic project), it follows that a symbiosis of life/machine naturally evolves so that intelligent life can move on in the cosmos, if and when the time comes that a planet supporting intelligent life is threatened with extinction. Is this what Plato meant by a moving image of eternity? And does this mean that life is passed on forever because it is part of this cyclical moving image of eternity?

Despite the fact that computers offer such novel approaches to technology and despite the fact that I believe a technologically sophisticated peoples may have already achieved a symbiosis of human/machine on other planets, I do not believe that the computer's digital logic can take us beyond the reach of our instinctive symbolic spiritual drive, because this drive is part of the law that rules our cosmos. Indeed, computers can create the neutral ground needed for the merger of imagination and digital-logic, but they will never take human imagination beyond what is already programmed in our genome by the cosmic instinct. We are, after all, limited by and ruled by our instincts, and our instincts for myth and technology are designed to take us off this planet—to send us soaring into space to discover other planets but only in the four-dimensional cosmos we live in. Whether we will ever break through into another unknown spacetime dimension, like Randall and other physicists suggest, remains deeply embedded ("rolled up," as Randall would say) in our DNA's secrets.

Revisiting Vallée, Jung: UFO as Spiritual Symbol

Although we have examined the UFO symbol as a technological inspiration, we must now consider it as a spiritual symbol that both Jung and Vallée, in different ways, claim that it is. We must also consider it in the light of its complementary real-life technological product, the spacecraft. Together, the UFO symbol and its evolved intellectual material counterpart, the present-day spacecraft, give rise to new horizons of discoveries in a manner that might even have surprised Heidegger. Spacecrafts are the result of an unrelenting continued manifestation of the UFO symbol to receptive people throughout the ages. Spacecrafts literally take our imaginations out of this world, but they may well turn out to be the *limiting* vehicles that inspire us imaginatively to go beyond their materiality and transport our bodies holographically to other worlds in the cosmos. In effect, they could end up being what inspires us to transcend them and forces us to crossover into a fifth dimension, a different reality in another way.

In the first chapter, Jacques Vallée pointed out that there appeared to be a symbolic law underlying the UFO phenomenon.[295] In his view, the law has something to do with symbolic (spiritual) messages being sent to us from another dimension; this he saw as an underlying spiritual control system. To reiterate his view, *"UFOs are physical manifestations that simply cannot be understood apart from their psychic and symbolic reality. What we see in this symbol has nothing to do with an alien invasion. It is a spiritual system that acts on humans and uses humans."*[296] As a spiritual system it could be designed by a terrible, superhuman monstrosity; a gathering of sages; or "the maddening simplicity of unattended clockwork"[297] to bring us into a new dimension of reality. This reality would be one in which psyche and matter are related. Interestingly, John Mack, in reporting an abductee's experience, says the following:

> There is something almost organic about the met-
> aphors that run through his [the abductee's] case.
> The dark blob he is shown by the beings and the
> thread or string that brings him to the UFO seem to
> exist in a kind of gray area between thought and the
> physical world. Like waves and particles in quan-
> tum mechanics, they seem to be thoughts in one
> context and something physically real in another.
> They are not simply one *or* the other, thought *ver-*
> *sus* something physical, but rather are *both* depend-
> ing upon the context.[298]

Vallée's position on the subject of UFOs as a manifestation of a spiritual system that acts upon and uses humans cannot be over-emphasized. Again, Vallée states:

> *In such a physics, UFOs could come from earth without necessarily being human inventions, or they could come from another galaxy without necessarily being spacecraft* [emphasized in text].[299]

But why would a control system (let us assume it is control-ling us from some other galaxy or unknown dimension beyond our reach) produce or instill lawfulness in such an absurd, illogical, irrational, and mythological manner and still be a system that has to do with our developing spirituality or consciousness? How is it possible to appropriate a psychic manifestation of matter through such a theater of the absurd? Vallée does not answer this in any adequate way, possibly because the spacecrafts we produce con-flict with his position of a spiritual system.

People like mythologist Joseph Campbell convinced many in North America in the late 1980s and early 1990s, during a TV series with Bill Moyers, that myths give witness to the fact that we are all inter-related *on earth*. According to Campbell, all aspects of myth and its rich cache of archetypal characters are imagined, purely psychic phenomenon, which for some unknown reason hap-pen to underlie and motivate our religious psyches. Campbell sug-gested that because the mythologies of the earth were universally structured, they were the result of nothing more than a common universal imaginative archetypal phenomenon operative in human-kind. Although he did not identify the phenomenon as instinctive and biological or as cosmically propagated, he did identify it as spiritual in nature, working through stories of virgin births, ini-tiation rituals, human sacrifice, the imaginative creation stories of sky gods descending to earth, and so forth. Yet given the fact that our own technology can now produce spacecraft not unlike these vehicles of our fantasies, how is it possible to continue to relegate this phenomenon to an ethereal imagination? Hence, a concomitant question is, are these once-proclaimed fantasies becoming manifest realities for a reason? Is there now a greater, unidentified urgency underlining the need to acknowledge their reality?

As Jung suggested, are these UFO symbols projecting materially onto consciousness for a reason that is not just to manifest the psychic wholeness that the mandela symbol represents but to orchestrate the technological results we see all about us on earth? Has the UFO symbol faithfully been revealing humanity's destiny over the ages, patiently stoking our technological instincts to maturity? Is this constant bombardment by the technological instinct that is manifesting itself in these UFO symbols an attempt to create a new awareness in us of just what spacetime represents—to usher in a new dawn? Have we, at the "end of philosophy," finally arrived at the point where we can begin to penetrate the mystery that underlies what I believe to be a message from the cosmos?

If this assault by UFO appearances is spiritually real, then should we continue to think of them as simply illusionary projections of no value, especially in the light of the technology we have achieved today? If UFOs are embedded in our instincts, is the consciousness that continues to think of them as illusionary itself deceived? If Jung is right and the unconscious materializes these UFOs, then a technologically sophisticated consciousness, capable of producing facsimiles of these projections, must begin to study this UFO phenomenon from scientific perspectives that includes its own biological makeup and not simply label them illusionary phenomenon. No longer can we retain the old-style attitude that the unconscious is the projector of illusion and that therefore the UFO phenomenon must continue to be considered as illusionary. Whichever way one chooses to interpret this symbol—as a spiritual or technological instinct—what holds true and what cannot be denied *is that technology is now producing a technological reality inspired by UFOs.* And because technology and the sciences are underlying this new reality, they need to reassess the value of the dynamics of breakthrough appearances.

In this respect, we can think of the UFO as a strange attractor, a symbol deeply embedded in our genome that has lead us to produce the spacecraft we know today. The Mandelbrot set, a much overused example, contains an uncanny reoccurring strange attractor that I will refer to in order to help make my point about instincts; in this case, about lawful technological instincts and their materialized products. The strange attractor, a parcel of intrinsic logic locked within the mesmerizing tendrils of a dramatic design, does not only occur in Mandelbrot's mathematical formula, but it also occurs in all aspects of nature, including chemical reactions and even in the DNA molecule. Surprisingly, the latter offers us an excellent visual picture of nonlinearity, replete with recurring embedded logic. The strange attractor in DNA occurs within a meandering sequence of base-pair couplings (CG, TA), which do so randomly but always with each other. The DNA code keeps popping up in the midst of "junk" DNA, as though to remind us that a remnant of an originally programmed logic will always remain intact. In the case of DNA, the strange attractor is the intrinsic logic that produces the human zygote; in the case of the recurring UFO symbol and its appearance at random to certain people, it is an absurd visual illogic that hides in a gene somewhere in our genomes. Yet this visual illogic inspires an intrinsic parcel of logic that I have been referring to as a technological instinct.

Instinctive Lawful Logic: the Cosmic Connection

Would this instinct for technology be possible if human beings were caught in a strictly Darwinian evolutionary scenario and only that? If evolution and only evolution was involved, how could the ancient mind have developed such a technologically biased imagination so quickly? Generally, we consider our forebears one step removed from primates. Julian Jaynes argued for a bicameral mind, one which he envisioned the ancient Greeks listening to through the auspices of their right hemispheres, in order to hear the commanding voices of their own left hemispheres—voices they thought came from the gods. Could such a bicameral mind—or some other version of an ancient mind—have developed these complex yet extremely creative and imaginative scientific notions of gods speaking to them from spacecrafts?[300] If so, why did this happen? Of course, our myths tell us that along with the spacecrafts were the biblical gods who descended to earth and/or ascended to heaven on the wings of what were generally referred to as chariots. In those times, they were described in rudimentary technological language. Today, these accounts of chariots are more readily recognizable to us as descriptions of spacecraft. How, then, can we not concede that all of our dominant, current religious beliefs are based on such visions of spacecrafts and encounters with gods and/or angels by our prophets? In other words, how was it that an imagination then could match our imagination now, in respect to describing UFOs? And why is it that appearances of the UFO strange attractor spaceships still rule the twenty-first-century imagination, perhaps more powerfully than ever before in history?

These technological instincts which, I propose, the human race has inherited from the "gods" are innate to us because of our cosmic connections or energies, and they give witness to our destiny as cosmic creatures. The instincts appear to have less to do with our limited earthly Darwinian evolution than with a universal cosmic biological inheritance. Indubitably, the appearances of UFO symbols fuel some people's desires to be like gods and/or to spend their eternal life with God in heaven. These images are in desperate need of reconsideration. Despite the fact that UFOs are meant to inspire our technological instincts, they also inspire our spiritual instincts. Hence, it would appear that we are *driven by a life instinct that must be passed along in the cosmos.* Our task, now more than ever, is to interpret the significance of the cosmic symbols we have been and are privy to, whether considered spiritual or cosmically lawful manifestations, and attempt to decipher how the instincts for technology or for the gods work in our genome and throughout the cosmos. It is these appearances that offer a very important clue as to how our destiny on earth and in the cosmos will evolve.

The more the technological instinct takes over on earth, the more a scientific consciousness takes hold to lead the way. The more a scientific consciousness conquers the mysteries underlying the physical laws of the cosmos, the more it will justify its search for other worlds and for other planets or galaxies on which to settle and from which to begin *in vitro* experiments to create new hybrid versions of intelligent life on these new worlds. This, then, is our instinctive connection to the intelligent principle in the cosmos: it dominates our instinct for propagating life. We cannot ignore the fact that our technology is making things happen today that more and more confirms and explains our ancient stories of what the earth was like when technocrat gods walked on it. Thus, I think it is safe to say that, because of our technological instincts, the more technologically sophisticated we become, the more we reinforce our so-called ancient mythical constructs that contain the story of our *essential* technological beginnings.

The cosmic instinct for technology will not necessarily accomplish this cosmic exercise of passing along life for the first time. Plato and the early Greeks indeed could have been right in this respect—there might well be an eternal return at work in the cosmos, with part of our present genetic endowment having been involved in this eternal return previously. Our ancient mythological stories ought no longer be ignored. They confirm Heidegger's ("Es gibt") and Emmanuel Levinas' ("il y a") claims that there is, indeed, an ancient voice out there that beckons "come" to us.[301] This ancient voice, not clearly identified by either Heidegger or Levinas, calls—whether they would admit it or not—from out of the cosmos.

The technological instinct speaks for itself, and more than this, it shows itself first in a quest for the merger of life with machine. It is the technological instinct that reveals itself to be the enabler of intelligent life; it is this instinct that passes along intelligent life in the cosmos. But the technological instinct would be nothing without symbolic direction, without the breakthrough of appearances. Scientists might not agree or admit to where their inspiration has always come from at this point, but I think it is safe to say that it comes from their connections to the cosmic instinct—the UFO symbol being among the prime instigator that connects us to this cosmic energy. To scientists, any symbolic inspiration that might tempt them to embrace nonlinear, "mystical-like" thinking is relegated to the realm of myth, yet more scientists end their books on a mystical note than any other way. Einstein was no exception; he had a phrase he liked to use in regard to his most mystical views of the cosmos' God. When he thought about the majesty of the laws of the cosmos, he described it as experiencing a "cosmic religious feeling."[302]

Scientific inspiration comes, indubitably, from our symbolic instincts, which are genetically programmed to tune into the spiritual laws of the cosmos and hence are readily receptive to the spiritual symbols that break through from the cosmos, whether within us or without us. The two—technology and myth—are closely related and connected by the strange attractor logic of DNA, be it through the appearances of UFOs, or aliens, or gods. These appearances are part of our cosmically inherited instinctive natures and, I believe, are always predestined to emerge at some point in time on any life-supporting planet, in any galaxy (perhaps even in a parallel universe, if they exist), in order that life continue its cyclical journey through the cosmos.

Is that all life is, then—part of a continuous repetitive cycle, even in macroscale? Until we break through the cyclic pattern's equilibrium, if we ever do, that is about all that can be said about it at this moment in time.

ENDNOTES

Introduction

1   See http://www.mcrobert.org

Chapter 1   Cosmos As a Psychic Control System

2   Jacques Vallée, *Dimensions* (New York: Contemporary Books, 1988).
3   Ibid., 285. (italics added)
4   Ibid., 289.
5   Jacques Vallée, *The Invisible College* (New York: E.P. Dutton, 1975).
6   Jacques Vallée, *Forbidden Science* (New York: Marlowe & Company, 1996), 431.
7   Jerome Clark, Exclusive Interview: Vallée Discusses UFO Control System, *FATE* magazine 1978, reprinted with permission by National Institute for Discovery Science. http://www.nidsci.org
8   Ibid., 290, (italics in text).
9   Vallée, *Forbidden Science*, 421.
10 Vallée, *Dimensions*, 71.
11 Ibid., 71.
12 Vallée, *Forbidden Science*, 420.
13 John E. Mack, *Abduction: Human Encounters with Aliens* (New York: Charles Scribner's Sons, 1994).
14 Ibid., 422.
15 Mack, *Abduction*, 413.
16 Vallée, *Forbidden Science*. This book constitutes a journal that Vallée kept between the years 1957 and 1969. What is remarkable about it is that

although still in his teens and early twenties, the ideas he germinated then were later explored and elaborated. For example, he wrote a novel *Dark Satellite* in 1961, where he invented a social system "that would replace both Marxism and capitalism (p.45)."

17   Paul David Pursglove, ed., *Zen in the Art of Close Encounters* (Berkeley, California: The New Being Project, 1995). The editor argues in favor of UFO being a koan-like experience.

18   Mack, *Abduction*, 420.

19   Michael Heim, *Virtual Realism* (New York: Oxford University Press, 1997), chapter 7.

20   Ibid., 145. I do not know whether Jung addressed the phenomenon of little gray men in respect to UFOs; he did, however, address the symbolism of UFOs.

21   Carl G. Jung, et al., *Man and His Symbols* (New York: Dell Publishing Co., Inc., 1964), 285.

22   C. G. Jung, *Memories, Dreams, Reflections* (New York: Vintage Books, 1965), 323.

23   Ibid., 324.

24   The dates vary depending on which school of scientists one belongs to.

25   Christopher Wills, *The Runaway Brain* (New York: Basic Books, 1993).

26   Carl G. Jung, CW 8, p.29.

27   Jung, CW 8, 223.

28   Jung, CW 9 i, 157.

29   Jung, CW 8, 157.

30   Jung, CW 8, 136.

31   Joseph F. Rycklak, *Introduction to Personality and Psychotherapy* (Boston: Houghton, Mifflin Company, 1973), 144–145.

32   Joland Jacobi, *Complex Archetype Symbol in the Psychology of C. G. Jung*(New York: Bollingen Foundation Inc., 1959), 76, 98.

33   Paul Ricoeur, *The Symbolism of Evil* (Boston: Beacon Press, 1967).

34   Immanuel Kant, *Critique of Pure Reason*, trans. Norman Kemp Smith (New York: St. Martin's Press, 1965). Kant lived his entire life (1724-1804) and died in Konigsberg.

35   Ibid., 65–66.

36   Jonathan Feng and Mark Trodden,"Dark Worlds", *Scientific American*, November, 2010, 40, 45. In this article on black holes Feng and Trodden speculate about what this matter might be made of and suggest "hidden worlds". "Could there be a hidden world that is an exact copy of ours, containing hidden versions of electrons and protons, which combine to form hidden atoms and molecules, which combine to form hidden planets, hidden stars and even hidden people? (43)"

Chapter 2 Holographic Dynamics: Poets, Mystics, Philosophers, and Physicists

37 For example, at least one hundred people spotted five mysterious UFOs hovering in the skies on a clear Saturday night above Stratford-upon-Avon on July 21, 2007.

38 Jacques Valle, *Dimensions: A Casebook of Alien Contact* (Chicago: Contemporary Books,1988), 22.

39 Arthur Zajonc, *Catching the Light* (New York: Oxford University Press, 1993). See chapter 8, 188, *passim*.

40 Immanuel Kant, *Dreams of a Spirit-Seer* (Bristol: Thoemmes Press, 1992), 3–33.

41 Ibid., Appendix III (remarks of the editor Frank Sewall),161.

42 Ibid., 7.

43 Ibid., 7.

44 Ibid., 4.

45 Kant, *Critique of Pure Reason*, 62.

46 Ibid., 49.

47 Ibid., 50.

48 Michael Talbot, *The Holographic Universe* (New York: HarperCollins Publishers, 1991). On the subject of how consciousness works in the brain, which includes how images work, see Christof Koch and Susan Greenfield's debate, "How Does Consciousness Happen? *Scientific American*, October 2007, 76-83.

49 Emmanuel Swedenborg, *Conversations with Angels*, ed. Leonard Fox and Donald L. Rose (Westchester, PA: Chrysalis Books, 1996), 17. See also Swedenborg's *Heaven and Hell*, 517.

50 Kant, *Dreams of a Spirit-Seer*, 4.

51 Ibid., 30.

52 Ibid., 124-6.

53 Ibid., 149-150.

54 *Scientific American*, October, 1996, Volume 275, No.4. See article by Hugh S. Lusted and R. Benjamin Knapp, "Controlling Computers with Neural Signals," 82.

55 Kant, *Dreams of a Spirit-Seer*, 142.

56 Ibid., 141.

57 Talbot, *The Holographic Universe*.

58 Recently, John L. Casti has argued in favor of a quantum reality of some sort as perhaps underlying reality. See *Scientific American*, October, 1996, "Confronting Science's Logical Limits," 102 ff. Casti argues that if we unshackled ourselves from the rigors of mathematical logic, we might be able to tackle ultimate questions in a reasonable way. He

states: "Our work strongly suggests that in the arts as well as in the natural sciences and mathematics, the human creative capacity is not subject to the rigid constraints of a computer's calculations. Penrose and other theorists have conjectured that human creativity stems from some still unknown mechanisms or rules, perhaps related to quantum mechanics. By uncovering these mechanisms and incorporating them into the scientific method, scientists may be able to solve some seemingly intractable problems," (105).

59   *Dreams of a Spirit-Seer*, 9-12.

60   *Dreams of a Spirit-Seer*, 150.

61   *Dreams of a Spirit-Seer*, Appendix IV, 162.

62   Michael Talbot, *The Holographic Universe* (New York: HarperCollins Publishers, 1991).

63   Talbot, *The Holographic Universe*, 15. See Figure 1.

64   *The Holographic Universe*, 35 *passim*.

65   *The Holographic Universe*, 35-37. The Einsteinian group decided to stump Bohr with the paradox that occurred when a positronium atom split into negative and positive charges. According to quantum physics, although these atoms travel in different directions and no matter how far apart they travel in space, they will always be found to have identical angles of *polarization*. Einstein's group proffered that an observer can only see one at a time and thus see only one wave and one particle, not two. Why, then, does the other unseen particle take on the same identical angle of *polarization*? Communication between them cannot travel faster than light, as this is limited by the special theory of relativity, so how could Bohr be right in his claims that only an observer could make the wave take on its particle form in the face of the paradox of the other particle's simultaneous *polarization*? Bohr, unperturbed by this Einstein-Podolsky-Rosen paradox, explained that their paradox was based on the fact that they were viewing the particles as separate entities, instead of as part of an indivisible system.

66   Ibid., 138.

67   Corey S. Powell and Madhusree Mukerjee, "Cosmic Puffery," *Scientific American*, September 1996, 22. Currently, cosmologists, astrologers, and others in closely related fields of study are stymied by new evidence that is contradicting their theories of the age of the universe. What was thought to be a much older universe is apparently turning out to be a much younger universe, estimated now to be about nine to twelve billion years old—in fact, young enough that scientists can barely fit in the ancient stars. (This is a very contentious issue; at the moment in cosmology, the generally accepted age of the universe is fourteen billion years.) To throw even more confusion into the picture, scientists

at the University of Edinburgh have observed galaxies in the very early universe's formation that are oddly very old in appearance. Much of the consternation has to do with establishing whether the universe is expanding or contracting and determining whether there is a neat formula that can encompass inflationary cosmology with particle physics. It seems to me that this quandary has to do with observing time past and present time that meet through interference waves and produce a holographic image. Differences in time can be explained more easily from a holographic perspective, where past, present, and future are interactive simultaneously than it can be from limited scientific perspectives designed to be exclusive of each other.

68   Talbot, *The Holographic Universe*, 54-55.
69   Ibid., see ch. 1, "The Brain as Hologram," and ch.2, "The Cosmos as Hologram."
70   Ibid., 258.
71   Ibid., 287.
72   Talbot, *The Holographic Universe*, 291.
73   Mark C. Taylor, *Altarity* (Chicago: The University of Chicago Press, 1987). Taylor describes their dynamics and those of others such as G. W. F. Hegel, Martin Heidegger, Emmanuel Levinas, and Soren Kierkegaard.
74   C. G. Jung, *Memories, Dreams, Reflections*, ed. Aniela Jaffe (New York: Vintage Books, 1973), 323. See chapter 2 for full quote.
75   Chris Scott, "The soul departs, the brain arrives," *Globe and Mail*, August 5, 2006, section D4. The article reviews a book by Raymond Martin and John Barresi, *The Rise and Fall of Soul and Self: An Intellectual History of Personal Identity* (New York: Columbia University Press, 2006). James Lovelock's *Ages of Gaia* is published by W. W. Norton, New York, revised and updated in 1995.

Chapter 3   Cosmic Genes and Holographic Projections

76   Richard Dawkins, *The Blind Watchmaker* (London: Penguin Books Ltd., 1986).
77   Ibid., 244.
78   Ibid., 207-208.
79   Ibid., 215.
80   Using Dawkins theory, we could add to this, if we wanted to stretch an inverse point, that a modern female generally projects her "undeveloped animus gene" on a similarly undeveloped anima gene in the male. In thus choosing for the intellectual male over other types of males, she reinforces, by virtue of her undeveloped animus, a false sense of intelligence onto the male and thus helps to create an irrational or illusionary worldview.

In other words, this reinforcement of the intellectual-ego of the male, in turn forces him to respond by becoming even more pseudo-intellectual and more creative or, if you like, even more destructive and irrational in the kind of world he builds to live up to this female projected immature animus image upon him. That all this anima-animus projecting is genetically programmed is argued by Jung's psychology in respect to female-male interaction. The whole scenario works conversely, too, with males' genes choosing for the perfect anima image, one that in the past and even today has managed to keep most women in shackles, as they (the women) choose to but cannot live up to the male's projection.

81   Dawkins, *The Blind Watchmaker*, 66ff. Biomorph-land is where Dawkins' computer "critters" live.

82   Ibid., 195. Analogies in the hands of "cranks" are dangerous. If used at all, they must be carefully used by a discerning, inspired scientist.

83   James Gleick, *Chaos: Making a New Science* (New York: Penguin Books, 1987), 150. This is the now famous early image of the butterfly that Edward Lorenz, a meteorologist, stumbled upon while searching for a way to predict long-term weather forecasts. Feeding data that consisted of three co-ordinates into the computer, Lorenz discovered that when he accidentally eliminated what seemed to be inconsequential decimal data by rounding off a decimal point to three rather than six digits, the computer printout changed in a radical way. He had chanced to factor in an infinitesimal error that produced an entirely new weather forecast, which was not even remotely connected to the previous one. Eventually, he was to discover that when certain formulas (he used twelve) were programmed to repeat themselves at regular points, the plotting of these points stayed within certain parameters (orbits or paths), never intersected themselves, never ran off the page, and traced a strange distinctive shape that consisted of a double-spiral in three dimensions that manifested itself as butterfly wings or owl's eyes. He had produced the image of a strange attractor in phase space, an image that is now called the Lorenz attractor.

84   Dawkins, *The Blind Watchmaker*, 57.

85   Ibid., 64.

86   Ibid., 60. "I still cannot conceal from you my feeling of exultation as I first watched these exquisite creatures emerging before my eyes. I distinctly heard the triumphal opening chords of 'Also sprach Zarathustra' (the *2001* theme) in my mind. I couldn't eat, and that night 'my' insects swarmed behind my eyelids as I tried to sleep." See also p. 58, p. 61, for diagrams of his results, plus pp. 62ff.

87   Ibid., 63.

88   Ibid., 66ff.

89   David Michael Levin, *The Opening of Vision* (New York: Routledge, Chapman & Hall, Inc., 1988), 473ff.

90 Ibid., 477. (italics added)

91 Ibid., 482.

92 James Gleick, *Chaos*, citing Edward Lorenz's, "The Butterfly Effect," 30.

93 Ibid., 95ff.

94 In 2010 it was discovered that Neanderthal genes survive in almost everyone outside of Africa and make up to 4% of our genomes. The Wall Street Journal, May 6, 2010. See also, http://fora.tv/2011/01/18/ Dr Jean-Jacques Hublin, Neanderthals Deciphered (accessed September 23, 2011)

95 Tabitha M. Powledge and Mark Rose, "The Great DNA Hunt," *Archaeology*, September/October 1996, 36ff.

96 Christian de Duve, *Vital Dust* (New York: BasicBooks, 1995).

97 *The Gazette*, August 7, 1996. The co-authors of the study are David McKay and Everett Gibson of the Johnson Space Center in Houston; Kathie Thomas-Keprta of Lockheed Martin, a NASA contractor in Houston; Hojatollah Vali of McGill University; Christopher Romanek of the University of Georgia laboratory in Aiken, S.C.; and Simon Clemett, Xavier Chillier, Claude Maechlin and Richard Zare of Stanford University in California. More information can be found at http://www. marsnews.com/focus/life (accessed May 26, 2011).

98 J. Madeleine Nash, "Was the Cosmos Seeded with Life?" *Time*, August 19, 1996, 44. This issue has a more comprehensive article titled "Life on Mars" by Leon Jaroff, 40ff.

99 *The Gazette*, August 14, 1996, reported by John Noble Wilford of the *New York Times*.

100 Mircea Eliade, *Images and Symbols* (Princeton, New Jersey: Princeton University Press, 1991), 76.

101 Richard Dawkins, *The Selfish Gene* (New York: Oxford University Press, 1976), revised edition, 1989.

102 Daniel C. Dennett, *Darwin's Dangerous Idea* (New York: Simon & Schuster, 1995).

103 *The Blind Watchmaker*, 216.

104 Richard Dawkins, *The Extended Phenotype: The Gene as the Unit of Section* (San Francisco: Freeman, 1982), 112. (italics added)

105 Joseph Campbell, *Masks of God: Oriental Mythology* (New York: Penguin Books, 1962/1972), in a foreword to the book titled "On completion of *The Masks of God*."

106 *Vital Dust*, 297.

107 Robert Wright, "Can Machines Think?" *Time*, April 1, 1996. Apparently, there is a new school of thinkers who call themselves "mysterians." Rutgers University philosopher Colin McGinn, author of *The Problem of Consciousness*, and David Chalmers, a professor of philosophy at

the University of California at Santa Cruz, who wrote the book *The Conscious Mind* (New York: Oxford University Press, 1996), argue that mystery underlies consciousness and the cosmos. They are very unlike Daniel Dennett, who in his book *Consciousness Explained* professes to have explained everything there is to know about consciousness. It is brain-only gray matter; that's all there is to it.

108 Antonio Damasio, *Descartes' Error* (New York: G. P. Putman's Sons, 1994). See also my "A Review with Reflections and Insights On: Descartes' Error" in *Explorations: Journal for Adventurous Thought*, Spring, 1995, Vol. 13, No.3.

109 While orbiting the earth, a Japanese scientist threw a boomerang out into space on March 19, 2008. As it would on earth, it came back to him. See http;//www.physorg.com/news125297819.html (accessed May 28, 2011).

110 The Blind Watchmaker, 315.

111 Ibid.

112 A planet named HD 189733b, some 63 million light-years away from our solar system, has been found to contain carbon-based molecules. Although it is too hot to favor conditions suitable for life as we know it, it constitutes a milestone in astronomers' hunt for extraterrestrial life in the cosmos. (Reported in the *Gazette*, Montreal, Saturday, March 22, 2008.)

113 *Holographic Universe*, pp.24 ff. Phantom-limb sensations and how we could construct a world out of the same kind of imaginary sensations are discussed.

114 Mircea Eliade, *Images and Symbols* (Princeton, New Jersey: Princeton University Press, 1991), 91.

115 Paul David Pursglove, ed., *Zen in the Art of Close Encounters* (Berkeley, CA: The New Being Project, 1995), 215.

116 Ibid., 227.

117 Ibid., 324.

118 Francis Crick, *The Astonishing Hypothesis* (New York: Charles Scribner's Sons, 1994), chapter 4, "The Psychology of Vision."

119 Norma J. Milanovich, *We the Arcturians* (Albuquerque, New Mexico: Athena Publishing, 1990, 1995). One book I came across accidently has to do with describing a channeling experience that a woman wrote, with the co-operation of her two friends. Channeling means being the receptor of messages from extraterrestrials and is a very popular idea today. The problem with the book is that it is so transparently unsophisticated. The authors can describe only very simple physical notions, constantly alerting the reader that their "beings" tell them anything more than this would be too complicated for earthlings to understand. It is as though the woman writing the book cannot expand on more complex scientific paradigms.

Chapter 4 Creating New Mythological Spaces

120 See http://www.neuroskills.com/tbi/btemporl.shtml for a description of the auditory and visual input in the temporal lobe of the brain (accessed May 26, 2011).

121 Semir Zeki, *Inner Vision* (London: Oxford University Press, 1999), 199ff.

122 Ibid., 202. Zeki states "that artists are neurologists, studying the organisation of the visual brain with techniques unique to them and their work, [that] when exploited scientifically, uncovers laws of cerebral organisation which scientists were previously ignorant of."

123 Dale Purves and R. Beau Lotto, *Why We See What We Do: An Empirical Theory of Vision* (Sunderland, Conn: Sinauer Associates, Inc., 2002).

124 Dennis Meredith, *Tricking the Eye or Trapping a Reflex? Duke* magazine, July-August,2000. See, http://www.dukemagazine.duke.edu/alumni/dm29/purves.html (accessed May 26, 2011).

125 Nikos K. Logothetis, "The Hidden Mind," *Scientific American,* August, 2002. See "Vision: a window on consciousness," 18-25.

126 Dale Purves and Catherine Howe, *Perceiving Geometry: Geometric Illusions Explained by Natural Scene Statistics* (New York: Springer, 2005). See also http://dukenews.duke.edu/2005/09/Purvesgeometrybook.html accessed May 26, 2011).

127 Josie Glausiusz, "How to Build an Invisibility Cloak", *Discover,* November 2006, 53-58.

128 Consider the fact that only a small percentage of military that fought in the Iraqi war had field experience; the rest had been trained through virtual reality re-enactments. See Howard Rheingold, *Virtual Reality,* (New York: Touchstone Simon & Schuster, 1991), and Kevin Kelly, *Out of Control* (New York: Addison-Wesley Publishing Company, 1994).

129 Charles Platt, "They Render Unto Bill," *Wired,* July 1996, 114ff.

130 129. Laurie McRobert, *Char Davies' Immersive Virtual Art and the Essence of Spatiality* (Toronto: University of Toronto Press, 2006). See also my essay, "Immersive Art and Technological Essence," *Explorations: Journal for Adventurous Thought,* Fall, 1996, Vol.15, No.1.

131 *Discover,* Vol. 17, No. 7, July 1996. See the "Breakthroughs" section on medicine, 30.

132 Vallée, *Dimensions,* 22.

133 Daniel Dennett, "Memes and the Exploitations of Imagination," *Journal of Aesthetics and Art Criticism,* Spring, 1990, 48, 127-35. He uses the example of the song "It Takes Two to Tango." This article can be found on the Internet at http://www.tufts.edu/as/congstud/papers/memeimag.htm.

134 Robert Doran, S. J., *Subject and Psyche: Ricoeur, Jung and the Search for Foundations* (Washington DC: University Press of America, 1997).

135 "Phantom Projections," *Discover*, July 1996, 90.

136 Julian Jaynes, *The Origin of Consciousness in the Breakdown of the Bicameral Mind* (Boston: Houghton Mifflin, 1976).

137 Bill Douthitt, *National Geographic*, December 2006, 38-57. The images for the article were provided by NASA, JPL, and Space Science Institute.

138 Adam Frank, *Discover*, April 2008. See his article "The Day Before Genesis," 54-60, which introduces three radical theories about how the universe began.

139 The Allan Institute for Brain Science is currently mapping the brain's genes. For more information, see http://www.brainatlas.org or http://brain-map.org (accessed May 26, 2011).

140 Michael Krantz, "Vulture Meat," *Time*, July 1, 1996, 34ff. About young entrepreneurs who compete with others in a Silicon Valley setup conference for the purpose of attracting financiers.

141 Howard Rheingold, *Virtual Reality* (New York: Touchstone, Simon & Schuster, 1991) 264. Rheingold reports his experience in Dr. Tachi's virtual reality setup, where he encounters himself: "He looked like me, and abstractedly I understood that he *was* me, but I know who me is, and me is *here*. He, on the other hand, was *there*. It doesn't take a high degree of sensory verisimilitude to create a sense of remote presence. The fact that the goniometer and the control computer made for very close coupling between my movements and the robot's movements was more important than high-resolution video or 3-D audio. It was an out-of-the-body experience, no doubt about it."

142 William Bramley, *The Gods of Eden* (New York: Avon Books, 1993), 306ff.

143 See www.news.bbc.co.uk/1/hi/education/582475.stm, an article titled "Virtual Teachers Could Enter Classroom," (accessible but only as a cached article, May 28, 2011).

144 On November 7, 2007, as I was writing this chapter, scientists announced that a new planet, within an inhabitable zone, had been discovered orbiting around a sun-like star. Known as Cancri, in the Cancer constellation, it has five planets orbiting around it.

145 Bramley, *The Gods of Eden*, 306ff.

146 Zecharia Sitchin, *The Wars of Gods and Men* (New York: Avon Books, 1985), 312ff. See also, Genesis 19:24-28.

147 Deepak Chopra, *Return of Merlin* (New York: Harmony Books, 1995).

148 Mircea Eliade, *The Sacred and the Profane* (New York: Harcourt Trade Publishers, 2001).

149 Dany Evanishen, ed. *The Raspberry Hut* (Summerland, BC: Ethnic Enterprises, 1994). The fairytale is titled "The Flying Ship," p.57 ff.

150 Laurie McRobert, "Great Walls and Celestial Iconics," *Explorations Journal for Adventurous Thought,* College Press, Spring 1998, Vol. 16, No. 3.

151 Jacques Vallée, *Dimensions* (New York: Contemporary Books, 1988).

Chapter 5 Spacetime Biologized

152 Christian de Duve, *Vital Dust* (New York: BasicBooks, 1995). More recently, NASA scientists have replicated how life might have begun and found clues that life began in deep space. See http://nai.arc.nasa.gov/news_stories/news_detail.cfm?ID=207 (accessed May 26, 2011).

153 Jocelyn Rice, *Discover*, March 2008, p.12, reports that Raoul Kopelman, who invented a tiny voltmeter to measure electrical fields in a cell, discovered that rather than being electrically dormant, the rat brain cells, which he flooded with the devices, "detected fields as strong as 15 million volts per meter throughout."

154 J. E. Lovelock, *Gaia* (New York: Oxford University Press, 1987), first published 1979.

155 NASA's Gravity Probe B Mission tested Einstein's theory of gravity and hence the warping and fabric of spacetime with four ultra-precise, spherical gyroscopes. They released the analyzed data in 2007. The data collected verified Einstein's theories.

156 Gary Taubes, "The Gravity Probe," *Discover*, March 1997, 63 ff.

157 Gerhard Staguhn, *God's Laughter* (New York: Kodansha International, 1994). Throughout the book and particularly in the conclusion, Staguhn aptly points out just how mythological physicists' constructs are.

158 Peter Byrne, "The Many Worlds of Hugh Everett," *Scientific American*, December 2007, 98-105,

159 "Breakthroughs", *Discover*, October 1996, 18.

160 Tim Folger, "At the Speed of Light: What if Einstein was wrong," *Discover,* April 2003, 34.

161 Gerhard Staguhn, *God's Laughter*, 243.

162 Ibid., 242.

163 Ibid., 233.

164 Ibid., 243.

165 Fred Alan Wolf, *Parallel Universes* (New York: Simon and Schuster, 1988), chapter 25.

166 Evelyn Fox Keller, *A Feeling for the Organism* (New York, W. H. Freeman and Company, 1983), 195.

167 *Discover*, October 2005, 8.

168 The discoverers that X in women is not a silent chromosome are Laura Carrel, assistant professor of biochemistry and molecular biology at the Penn State College of Medicine, and Huntington Willard, head of the Institute for Genome Sciences and Policy at Duke University.

169 Researchers at McGill University, among others, are Moshe Szyf and Tom Hudson. Epigentics hopes to explain why there are differences between people who are genetically identical, like twins. They even suspect that the ability to switch genes on or off may offer an evolutionary advantage by speeding the normal process up. Anne McIlroy, *Globe and Mail*, March 11, 2006, A4, A5.

170 Barbara Oakley, *Evil Genes: Why Rome Fell, Hitler Rose, Enron Failed, and My Sister Stole My Mother's Boyfriend* (New York: Prometheus Books, 2007). Oakley's chapter 3, "Evil Genes," offers an excellent explanation of how genes work, why people can have double doses of certain genes or alleles, and the havoc this might cause in the brain.

171 Dawson Church, *The Genie in Your Genes* (Santa Rosa, CA: Elite Books, 2007), 169ff.

172 For more information on lightning strikes, see http://tigger.uic.edu/labs/lightninginjury/overview.htm, (accessed May 26, 2011).

173 "Plants Using Quantum Computers," *Discover*, January 2008, 45.

174 David H. Freedman, "The Mediocre Universe," Discover, February 1996, 71.

175 "The Mediocre Universe," *Discover*, op.cit., 73.

176 Jacques Vallée, *Dimensions* (New York: Contemporary Books, 1988), 22.

Chapter 6 Science Fiction, Science and Evolving Imagination

177 Lawrence M. Krauss, *The Physics of Star Trek* (New York: HarperPerennial, 1995).

178 For a virtual reality field trip, see http://education.guardian.co.uk/evaluate/story/0,,1386927,00.html

For holographic teachers, see http://news.bbc.co.uk/2/hi/uk_news/education/582475.stm

(both sites accessed May 26, 2011).

179 On April 10, 1997, it was announced that photographs taken of Europa one of Jupiter's moons clearly indicates ice caps. Scientists now believe there is probably an ocean there and if an ocean therefore some form of life. Christian de Duve, Nobel laureate, has argued this thesis in his book *Vital Dust, Life as a Cosmic Imperative* (New York: BasicBooks, 1995).

180 Carl Sagan, *The Demon-Haunted World: Science as a Candle in the Dark* (New York: Random House, 1995).
181 Stephen Jay Gould, *Wonderful Life* (New York: W.W. Norton & Company, 1989).
182 Quantum Reality, 172.
183 The latest such experiment with virtual reality is a study associated with locomotor recovery in chronic stroke conducted by a group of doctors, of which Sung H. You, PT, PhD, assistant professor, doctor of physical therapy program, Hampton University, is one. See http://stroke.ahajournals.org/cgi/content/abstract/36/6/1166 (accessed May 26, 2011).
184 John E. Mack, *Abduction* (New York: Charles Scribner's Sons, 1994).
185 Lawrence M. Krauss and Robert J. Scherrer, "The End of Cosmology?" *Scientific American,* March 2008, 47-53.
186 Tom Thompson, "What's New? A glimpse at three technologies that could be the subsystems of tomorrow's desktop computers," *Byte,* April 1996.
187 Dawson Church, *The Genie in Your Genes: Epigenetic Medicine and the New Biology of Intention* (Santa Rosa, CA: Elite Books, 2007). See chapter 5, "The Connective Semiconducting Crystal," for a biological perspective of crystalline structure throughout the body's connective tissues, the brain, microtubules, etc.
188 Susskind, "Black Holes and the Information Paradox," *Scientific American*, April 1997, 52 ff.
189 "Black Holes and the Information Paradox," 55.
190 Ibid., 56, 57.
191 "Black Hole Feasts at Milky Way's Center," *Discover*, January, 2008, 46. Since Susskind posited his theory, we now know that when a black hole digests something as large as the planet Mercury, it leaves x-ray evidence of its dinner by displaying a formidable light show that we on earth can see for many decades to come through the powerful x-ray telescopes we now have.
192 "Memes" are the invention of Richard Dawkins and have been picked up by philosophers such as Daniel C. Dennett to explain how ideas proceed using a biological paradigm. See Dennett's *Darwin's Dangerous Idea* (New York: Simon & Schuster, 1995) and also, Richard Dawkins, *The Blind Watchmaker* (New York: Penguin Books, 1988), 158.
193 Jared Diamond, "Kinship with the Stars," *Discover*, May 1997. Diamond, in a tribute to Carl Sagan, points out just how ruthless Sagan's colleagues were to him by denying him membership in the Academy of Sciences. Diamond also reminds us that unless academicians speak to each other in a cross-disciplinary language when they write articles for nonspecializing journals, they will fail to make insights that naturally come from making analogies from one field to another.

194 "Black Holes and the Information Paradox," 56.

195 "Breakthroughs," *Discover*, December 1995, 26. A natural laser of this sort in space cuts through the background noise of the gaseous disk. Since it is six times brighter than the normal light of the glowing disk it provides a beam through which scientists can access its inner workings. This allows them to analyze information otherwise not available to them.

196 Will Hively, "X-ray Dreams," *Discover*,July 1995 70 ff. The discussion is about Charles Rhodes' X-ray laser microscope which he is attempting to develop in a practical way. Mostly, his theories on the subject have been proven to be correct. They are technically sophisticated, and therefore I recommend you read the article itself.

197 *Scientific American*, July 2005, 18, for a discussion on cosmology and the "flaw of averages."

198 Francoise Combes, "Ripples in a Galactic Pond," *Scientific American*, October 2005, 43.

199 Jeffrey Winters, "Cube Tube," *Discover*, December 1996.

200 Peter N. Sotts, "Spooky action at a distance," *Christian Science Monitor*. Verified by a Netscape Security Partner: VeriSign http://www.csmonitor.com/2001/1004/p15s1-stss.html (article no longer accessible).

201 V. V. Ivanov, et al., *Particles and Nuclei, Letters,*. No. 1 [116] 2003, 96-107.

202 Arthur Zajonc, *Catching the Light* (New York: Oxford University Press, 1995), 296 ff.

203 *Catching the Light*, 299.

204 Nick Herbert, "See Spot Run: A Simple Proof of Bell's Theorem," http://quantumtantra.com/bell2.html (accessed May 26, 2011).

205 Roger Penrose, *Shadows of the Mind* (New York: Oxford University Press, 1994), 337.

206 For example, consider the modern-day notion of spread spectrum—the spreading of a signal among many frequencies to prevent interference, eavesdropping, or jamming. It is interesting to note that Hedy Lamarr, an Austrian actress whose career hit its zenith in the 1940s, patented, along with a friend, composer George Antheil, the idea of using a piano rolls technique to jump from one frequency to the next in random order. Much like a pianist moves from note to note, so to the message jumps from one frequency to another. Lamarr and Antheil conceived this as a fine way to confuse the enemy during the Second World War. They saw devices similar to player-piano rolls as being installed shipboard and inside the torpedoes. This would protect and shield radio-controled torpedoes from German signal-jamming. The technique was not used until the Cuban Missile Crisis of 1962, when it was used for radio communications.

207 Zecharia Sitchin, *The 12th Planet* (New York: Avon Books), chapter 2, "The Sudden Civilization."

208 Anthony Aveni, *Empires of Time: Calendars, Clocks and Cultures* (New York, Basic Books, 1989).

Chapter 7 Reincarnation, Inherited Memory and the Genome

209 W. Y. Evans-Wentz, *The Tibetan Book of the Dead* (London: Oxford University Press, 1960), 178.

210 Jean M. Auel, *The Clan of the Cave Bear* (New York: Bantam Books, 1980).

211 Ibid., 28-29.

212 Christian de Duve, *Vital Dust: Life as a Cosmic Imperative* (New York: Basic Books, 1995).

213 Charles Siebert, "Unintelligent Design," *Discover*, March 2006, 32-39.

214 Steve Rose, *Lifelines: Biology Beyond Determinism* (New York: Oxford University Press, 1997), 270. See chapter 9 for an excellent narrative of how life may have begun.

215 Ibid., 252.

216 Robert M. Hazen, *Gen.e.sis: The Scientific Quest for Life's Origin* (Washington, DC: Joseph Henry Press, 2005), 235-240.

217 Ibid., 241-243.

218 De Duve, *Vital Dust*, 301.

219 Joel Achenbach, "The Origin of Life...Through Chemistry," *National Geographic*, March 2006, 31.

220 Graham Hancock, *Supernatural: Meetings with the Ancient Teachers of Mankind* (New York: Random House, 2005).

221 Ibid., 471-473. The reference Hancock gives is Francis Crick, *Life Itself: Its Origins and Nature* (London: Future Macdonald, 1982), 171-3.

222 Ibid. See Appendix III, 618. See also Rick Strassman, *DMT: The Spirit Molecule* (Rochester, Vermont: Park Street Press, 2001).

223 De Duve, *Vital Dust*, 15.

224 BBC NEWS/Science/Nature, "Meteorites carry ancient carbon," http://news.bbc.co.uk/2/hi/science/nature/4973696.stm (accessed May 26, 2011). No longer do we have to rely on cosmic dust. The Carnegie Institute of Washington study suggests that we can find samples of cosmic material in interplanetary rock and pieces of metal that land on earth.

225 Fortunately, after his death, I kept, as a souvenir of his "art," one such drawing similar to that which Hancock reproduces in his book, and had it framed. It is of three little men complete with tasseled hats.

226 David Whitehouse, BBC News Online Science, Oct.11, 1999. http://news.bbc.co.uk/1/hi/sci/tech/471786.stm (accessed May 26, 2011). The images are blurry but recognizable, although it is possible that if more than 177 cells sensitive to light and dark were involved, the images might be clearer.

227 Zecharia Sitchin, *The Earth Chronicles V: When Time Began* (New York: Avon Books, 1993), 3-5. This is the fifth book of a series of chronicles on the *Enuma elish.*

228 Jean Bottéro, *Mesopotamia: Writing, Reasoning, and the Gods,* (Chicago: The University of Chicago Press, 1992), 220. Bottéro's interpretation is theogonic, describing how the great gods evolved instead of the planets. Here, Marduk defeats the great primordial Mother, the monstrous Tiamat, cutting her body apart to create the heavens and stars above, while leaving the lower part of it intact to become Earth.

229 Sitchin, *The Earth Chronicles V*, 6.

230 Scientists are quick to point out that this kind of discovery does not necessarily imply that life exists. There are several ingredients necessary for life to emerge, including a stable heat source and the right chemical recipe. The picture is of geyser-like eruptions of ice particles and water vapor that appear to be coming from underground reservoirs on the south pole of Enceladus. From a news release of the Associated Press by Alicia Chang, which appeared in the *Globe and Mail* on March 10, 2006.

231 The crash occurred on October 9, 2009, with results announced on November 13, 2009. For more details on this mission see http://news. bbc.co.uk/2/hi/science/nature/8359744.stm (accessed May 26, 2011).

232 Zecharia Sitchin, *Book One of the Earth Chronicles: The 12th Planet* (New York: Avon Books, 1976).

233 Zecharia Sitchin, *Genesis Revisited* (New York: Avon Books, 1990), 160-162.

234 This necklace was found near Lake Titicaca. See http://www.msnbc. msn.com/id/23887855/
(accessed May 26, 2011).

235 Graham Hancock, *Supernatural*, 479.

236 De Duve, *Vital Dust*, 6, 20.

237 Robert Sapolsky, "The 2% Difference," *Discover*, April 2006, 42.

238 Semir Zeki, *Inner Vision* (London: Oxford University Press, 1999). See chapter seven, 58-69.

239 Antonio Damasio, *Descartes' Error: Emotion, Reason, and the Human Brain* (New York: G.P. Putman's Sons, 1994); see also Antonio Damasio, *The Feeling of What Happens: Body and Emotion in the Making of Consciousness* (New York: Harcourt Brace & Company, 1999).

240 Immanuel Kant, *Critique of Pure Reason*, trans. Norman Kemp Smith (New York: St. Martin's Press, 1965), 93. Emphasis added.

241 Mark Gerstein and Deyou Zheng, "The Real Life of Pseudogenes," *Scientific American*, 49-55.

242 Mark Simpson, "UFO Study Finds No Sign of Aliens," BBC News, Sunday May 7, 2006. The article can be found at http://news.bbc.

co.uk/2/hi/uk_news/4981720.stm (accessed May 26, 2011). Apparently the study was stamped "Secret: UK Eyes Only" and was made public after Dr. David Clarke of Sheffield Hallam University requested it under the Freedom of Information Act.

243 Lisa Randall, *Warped Passages: Unravelling the Mysteries of the Universe's Hidden Dimensions* (New York: HarperCollins Publishers, 2005), 3.

244 Ibid., 49.

245 Laurie McRobert, *Char Davies' Immersive Virtual Art and the Essence of Spatiality* (Toronto: University of Toronto Press, 2006).

246 Dean Hamer, *The God Gene: How Faith Is Hardwired into Our Genes* (New York: Random House, Inc., 1994).

247 Ibid., 73.

248 Eugene d'Aquili and Andrew B. Newberg, *The Mystical Mind* (Minneapolis: Fortress Press, 1999).

249 Ibid., 132.

250 Ibid., 130.

251 Larry R. Squire and Eric R. Kandel, *Memory: From Mind to Molecules* (New York: Scientific American Library, 1999). To get anywhere near this subject, we would need to begin with molecular biology and cognitive processes, and there is not too much available in this area. Neuroscientist Larry Squire and neurobiologist Eric Kandel, who examine molecular biology and cognitive processes, hope someday to take this one step further by exploring the biology of memory.

252 Steve Rose, *The Making of Memory* (New York: Anchor Books, Doubleday, 1992). Stephen Rose's research in this area has revealed that a five-amino-acids-long peptide appears able to rescue memory loss by identifying the cell adhesion molecules needed to make a new class of proteins. Apparently, each time a long-term memory must be stored, a new class of protein must be made to assure that it adheres in the proper configuration. If the wrong configuration is achieved by improper molecular adhesion or if antibodies are introduced, long-term memory cannot be made.

253 Daniel C. Dennett, *Breaking the Spell: Religion as a Natural Phenomenon* (New York: Viking, 2006).

254 Steven Pinker, *How the Mind Works* (New York: W. W. Norton & Company Inc., 1997).

255 Steven Rose, *Lifelines: Biology Beyond Determinism* (New York: Oxford University Press, 1997). For an online debate between Pinker and Rose, see http://www.edge.org/3rd_culture/pinker_rose/pinker_rose_p4.html (accessed May 26, 2011).

256 Newberg and d'Aquili, *The Mystical Mind*, 129.

257 The Haplotype Project is one example of a collaboration of scientists working together throughout the world to examine differences in heritable blocks of DNA. This remains a thorny ethical and political issue, as racist groups use the Internet-posted information to fuel their agendas. However, the scientists involved in the project believe that such Haplotype mapping can eventually lead to the discovery of new ways to fight inherited disorders.

Chapter 8   Rolled-Up Dimensions

258 Lisa Randall, *Warped Passages*, 454.
259 Ibid., 455. Randall's quote is slightly dated, since quantum teleportation is more than theory. The teleportation of "key properties from one particle to another without a physical link" has been achieved by a team of physicists who teleported light across the Danube River. http://news.bbc.co.uk/1/hi/sci/tech/3576594.stm (accessed May 26, 2011).
260 Lee Smolin, *The Trouble with Physics: The Rise of String Theory, the Fall of a Science, and What Comes Next* (New York: Houghton Mifflin, 2006). Smolin argues that physicists might have come to the end of their "strings" and entered the realm of fantasy, as there has been no empirical evidence to back them for over sixteen years now.
261 M. G. Lord, "Impossible Journey?" *Discover*, June 2006, 38-45. Lord questions whether we are trapped on earth because of the deadly effect of cosmic rays on the human body, long term.
262 Laurie McRobert, "Great Walls and Celestial Iconics," *Explorations Journal for Adventurous Thought*, College Press, Spring 1998, Vol.16, No.3. See also http://www.mcrobert.org.
263 These new results were presented by Princeton University at a news conference on March 23, 2006. A space-borne instrument called the Wilkinson Microwave Anistropy Probe (WMAP) was launched by NASA in 2001. The physicists involved were Charles Bennett of Johns Hopkins University and Lyman Page and David Spergel of Princeton.
264 Stephen Jay Gould, *Wonderful Life* (New York: W. W. Norton & Company, Inc., 1989).
265 Lisa Randall, *Warped Passages*. Randall describes the LHC as "A high-energy particle collider that will bang together 7 TeV proton beams and produce particles with mass up to a few TeV," 464. She defines TeV (teraelectronvolt) as "a unit of energy equal to one trillion eV," 469. An eV (electronvolt) is "the energy required to move an electron against a potential difference of 1 volt," 462.
266 Ibid., 41.
267 Corey S. Powell, ed., "Data: Are We All Asians," *Discover*, Science, Technology, and the Future, May 2005, 12-13.

268 Evan Scannapieco, Patrick Petitjean, and Tom Broadhurst in an article (published in *Scientific American*, October 2002) titled "The Emptiest Places," and *Scientific American, Majestic Universe*, 2004 (a special issue), 37. The authors state that radio studies reveal diffused magnetized gas in several nearby galaxy clusters. This implies that the Intergalactic Medium (IGM) as a whole is magnetized.

269 Ibid., 39.

270 Ronald J. Reynolds, "The Gas between the Stars," *Scientific American, Majestic Universe*, 2004 (a special issue), 47. First published in *Scientific American* in January 2002.

271 Chris Quigg, "The Coming Revolutions in Particle Physics," *Scientific American*, February 2008, 46-53. Contains a comprehensive explanation of what scientists believe the Higgs particle to be and what it is that might or might not happen when they run the Large Hadron Collider (LHC) in hopes of producing it. The LHC began colliding beams in November 2009 and is scheduled to reach full energy in 2014.

272 Lisa Randall, *Warped Passages, 352.*

273 On July 4, 2012, scientists at CERN announced to the world that they are confident that a Higgs-like particle (the God particle) has been discovered. This new discovery will revolutionize physics by allowing physicists to test what up until now were just theories.

274 Scannapieco, et al., "The Emptiest Places," *Scientific American*, 36.

275 Martin Mittelstaedt, "Does Power Corrupt," *Globe and Mail*, March 28, 2006, A3.

276 Joachim Wambsganss, "Gravity's Kaleidoscope," *Scientific American, Majestic Universe*, 2004 (a special issue), 49. This article explains and contains drawings of how gravitational lensing works in the cosmos. First published in *Scientific American* in November 2001.

277 Alana Coates, "Student's electrifying discovery has potential to change the world," *The Gazette*, Montreal, April 8, 2006, B2. See also http://www.physics.uwo.ca/teamcana/2004/madiraju_report.pdf.

278 Zeeya Merali, Splitting Time from Space," *Scientific American*, December 2009, 18-21. Petr Horava, a physicist with the University of California, Berkeley, reworks Newton's idea that space and time are not equivalent. Horava goes back to the early universe, where quantum gravity ruled and time was absolute.

279 Seth Shostak, "Drake's Brave Guess," *Discover*, Science, Technology, and the Future, May 2006, 58-61.

280 Joel Achenbach, "The God Particle," *National Geographic*, March 2008, 95-105. See also 101.

281 Robert Langs and Anthony Badalamenti with Lenore Thomson, *The Cosmic Circle* (New York: Alliance Publishing, Inc., 1996).

282 Ibid., 137.

283 Ibid., 126.
284 Ibid., 180.
285 Anthony Aveni, *Empires of Time: Calenders, Clocks, and Cultures* (New York: Basic Books, 1989).

Chapter 9   Appearances and the Technological Instinct

286 Laurie McRobert, *Char Davis' Virtual Immersive Art and the Essence of Spatiality* (Toronto: University of Toronto Press, 2007).
287 Martin Heidegger, Basic Writings, 316, 317.
288 From an excerpt titled "The Brain Behind the Brain," *Business Week*, July 17, 1995, 71.
289 Otis Port, "Computers that think are almost here," *Business Week*, July 17, 1995. I have quoted from excerpts on pages 68, 69. Although biological computing is a hot newcomer on the scene, made possible by the science of nanobiotechnology, it is still in its infancy at this writing (2011).
290 Ibid.
291 Aveni, *Empires of Time,* 249. Aveni states: "The idea of time going round and round, of events being conceived as marks on a circle thematically repeating themselves, is totally at odds with our way of thinking of time."
292 Martin Heidegger, Basic Writings, 303.
293 An article propounding Andrei Linde's ideas, "The Self-Reproducing Inflationary Universe," can be found in the November 1994 issue of *Scientific American.*
294 I refer the reader to Zecharia Sitchin's work on this subject. All six books are published by Avon Books, New York.
295 Jacques Vallée, *Dimensions* (New York: Contemporary Books, 1988).
296 Ibid., 284-285. (italics added).
297 Ibid., 280.
298 John E. Mack, *Abduction* (New York: Charles Scribner's Sons, 1994).
299 Ibid., 285.
300 Julian Jaynes, *The Origin of Consciousness in the Breakdown of the Bicameral Mind* (Boston: Houghton Mifflin Company, 1976).
301 Mark C. Taylor, *Altarity* (Chicago: The University of Chicago Press, 1987), 210-11.
302 Corey S. Powell, "My Three Einsteins," *Discover,* October 2006, 43-7. See 47.

INDEX

Abductees, 11-26, 55, 91, 104, 120-
2, 129, 136, 146, 152, 176, 182,
197
Achenbach, Joel, 151, 221, 225
Aliens, 3-26, 90, 99, 102-5, 122,
128-9, 139, 153, 164, 176, 187,
196-7, 207n.13,
223n 242
Angels, 21, 32-8, 70, 72, 99, 122,
155, 160, 182, 194, 204, 209n49
Appearances,
ancient, 99; Angels, 182;
archetypal instincts, 15, 104;
biological nature of, 164-4, 172;
breakthrough, 107; circularity
of, 152, 162, 167, 176, 186-7;
cosmic related, 3, 16, 43,
77, 106, 106, 165, 186;
Dawkins, 51; DNA, x, 5, 19, 23,
24-5, 49-50, 62, 95, 106, 111,
114-122, 130, 135, 141, 148-9,
151-154, 161-8, 174-6, 187, 190,
203; dreams and, 2, 13, 15, 18,
39, 45-7, 77, 97, 116, 157, 161-3,
169, 178, 209n40, 208n22,
220n196; electromagnetic interferences and, 77, 165, 180-1;

four-dimensional, 176; genome;
genes, 9, 96, 156, 171, 176,
183, 205, see also DNA; gods,
193-5, 202-6; heavenly bodies, 47; holographic nature, 40,
117, 180; in passim, x, 1-12, 27,
56, 80, 97, 141, 170, 176, 189,
196, 203, 206; imagination,
22; Jesus, 72-3; Jung, 16; Kant,
23-5, 30-5, 39-40, 113, 163;
otherworldly projections, 20,
113, 121, 178; parallel universe,
130; projecting-receiving, 106;
Purves, 84, 87; real-time and
space, 105, 178; science, 39,
113, 162; spacetime, 108, 113,
123, 132, 133; spatial dimensions, 96; superimposing,
117-119; Swedenborg, 33-40;
technological instinct, 190-
1; technology, 92, 102; UFO
phenomenon, x, 3, 15, 20, 47-8,
72, 90-2, 99, 102, 130, 140, 190,
194, 197, 206, 213n95, 224n257;
worldview, 2, 56, 74, 101
Archetypes, 15, 17, 18, 51, 66-8, 74,
94, 167, 169; archetypal projec-